In Search of Flowers

PATRICK M. SYNGE
VMH

IN SEARCH OF FLOWERS

London
MICHAEL JOSEPH

First published
in Great Britain
BY MICHAEL JOSEPH
52 Bedford Square, London, W.C.1
1973

ISBN 0 7181 0872 8

Set and printed in Great Britain
by Tonbridge Printers Ltd,
Peach Hall Works, Tonbridge, Kent
in Baskerville ten on twelve point
and bound by
James Burn at Esher, Surrey

Old men ought to be explorers
Here and there does not matter
We must be still and still moving
Into another intensity
. . .

T. S. Eliot
East Coker, FOUR QUARTETS
(*Reprinted by permission of Faber and
Faber Ltd*)

*To my wife and the other patient friends who have
accompanied me on flower hunting trips and made
them so much more pleasant and rewarding.*

Contents

List of Illustrations

9

Introduction

The search for flowers in wild places has always enthralled me and in the following pages I have tried to describe some of those that I have seen. Of course there were many more. This is only a selection. There are very many more parts of the world that I have not visited, many valleys and ridges unknown even in the mountains I have visited. Each may have slightly different plants. This is part of the excitement. Also, at different seasons one may see different ones again. I have no complete knowledge of the flora of any one country such as can be obtained by residence, observation and collecting there over a number of seasons. However, I hope that this little book may be a help to other travellers, since it is possible without great difficulty today to visit most of the areas described. When I have had a choice I have nearly always tended to go to mountain areas; this is a personal preference as they are the areas, usually, which give me the greatest release and the most interest, but an equal number of interesting plants could be seen without ascending more than a few hundred feet in other parts of the world.

However, I have included the main and most prominent plants likely to be seen by the visitor who is interested in the areas of the Alps and Pyrenees and the eastern Mediterranean. Many of these are still abundant but some are becoming much less frequent while a few have never been common. All gardeners and travellers must now have a thought for conservation of the heritage of wild flowers that they so much enjoy, for the wild flowers of one region may be the garden flowers of another. Big game photography has replaced shooting to a very great extent in respect of our larger wild animals and has brought enjoyment and appreciation of them to a far wider range of travellers than ever before. The same is happening, although to a smaller extent, in the case of wild flowers. On a recent visit to Crete for instance, with a flower and bird tour, in one place we found only a single specimen of a choice fritillary. Thirty people photographed and enjoyed it and then left it undamaged to ripen and we hope spread its seed, and maybe to found a small colony. The same applies particularly to the terrestrial orchids both of the Alps and of the Mediterranean regions. They are wonderful and thrilling and in the case of the genus *Ophrys*, the genus of the bee and fly orchids, no two individuals seem to be exactly alike, so infinite is the variation. They may take ten or more years to flower from seed but they should be given their chance in their own habitat. Nor are they good or easy garden plants when transported home, practically all those so collected in the past have been lost within a few years. So it is surely far better to leave them where we find them and return just with a photograph for memory. In other cases where one finds an exceptional form of a common plant, such as the white form of *Cyclamen repandum*, one should exercise discretion. One should never take the only specimen of a rare plant, however tempting. The same is probably beginning to apply to some of the more visited regions of the Himalaya. In the case of shrubs or many alpines, or a specially good form of a wild heather, a few cuttings will often suffice for its introduction or even a pod of seed which can well be spared.

Even though one may know well a plant in the garden environment, it still gives a thrill and a special pleasure to see it growing in the wild though the specimens, perhaps, may not be nearly so fine. This may seem illogical to some people and be difficult to explain to the townsman. Nevertheless, it is undoubtedly a fact for the majority of gardeners and botanists, and long may it continue so. In spite of cries of pollution and desecration of many parts of the world, I am optimistic enough to believe that for many generations to come there will still be wild places and wild areas for us to enjoy, and probably the next generation will neither expect nor particularly want to see them in solitude although for a few this is likely always to be a desire. However, even in my lifetime, when I revisit popular areas in the Alps first seen as a boy many of the more striking wild flowers seem less abundant, while if one reads the enthusiastic descriptions of Reginald Farrer or other earlier writers on plants one realizes, even after making allowances for their ebullient writing, how much some of the flowers have declined. Many good areas and populations of plants are also lost through changes in the use of the land; areas are built over or damp areas drained thus causing a change in the flora, so there is constant change. This is inevitable and conservation has to be an active rather than a passive process as well as becoming a habit of mind.

To see specimens of our rarer garden plants and some that we regard as difficult in cultivation growing in the wild also helps us to understand the conditions which they are likely to require in gardens, the periods in which they are accustomed to grow and those in which they will be dormant. Apart from its interest it can be a great advantage towards their successful cultivation. However difficult a plant may have the reputation of being in cultivation, and however restricted and specialized the area from which it comes in the wild, I still believe that a sufficient understanding of its wild requirements should enable us to master its cultivation. There will always, however, be some challenges. Fortunately for gardeners, the range of tolerance of most plants is very wide and they can adapt themselves to growing successfully, under very different conditions, often better than in their native habitats.

On my various trips I have enjoyed the companionship of many friends and they have always been happy trips, as indeed such experiences should be. I am grateful to them all for their tolerance and patience and for the extra pleasure that they have given me. Soon I shall need to experience my trips through the eyes of a younger generation, and it is good to see so many flower lovers keen to travel and, indeed, accomplishing it very much more widely than formerly. I hope that this book may serve to whet their appetites for seeing plants in the wild and enjoying them, and may help to suggest particularly interesting areas and the best seasons in which to visit them.

Several of my companions have been kind enough to read sections of the book relating to areas in which we travelled and to give me helpful comments: Paul Furse on Turkey and Persia; Eliot Hodgkin on Spain, Corsica and the Julian Alps; Colville Barclay on Nepal, and I am most grateful to them. I also gladly acknowledge my debt to Juliet Tuck, my neighbour, for her patient and careful typing of my untidy manuscripts. Without this the book would have taken much longer to complete.

CHAPTER ONE

The Alps, Maritimes, Dolomites and Eastern Alps

THE ALPS

Visits to the Savoy Alps were my first mountain experiences and probably none have been more rewarding, both in the excitement and variety of the flowers seen as well as in their abundance. Since then I have been in the Pyrenees, the mountains of the Bernese Oberland and around Zermatt and Wengen in Switzerland, the Dolomites and the Julian Alps of northern Yugoslavia. Each area has its different flowers, its own endemics as well as many of the commoner flowering plants which are native to the whole range. I never tire of the thrill inspired by the first clumps of gentian, either *G. verna* or *G. acaulis*, after a spell of absence from the mountains. They are common but nevertheless are unique in their colour. The magic is always there, afresh for every visit. The scenery and the flora are so rich and varied, the hospitality and welcome of the hoteliers so excellent, the thrill of the sunset or the dawn from a mountain hut so magnificent, when often the peaks only are seen emerging above a sea of cloud; the freshness and crispness of the air such a tonic after work in a London office, that there can be few holidays to compare with it. Although since my early visits to the Alps, I have ranged further afield, in beauty of scenery they still compare favourably with other more exotic-sounding areas.

For the finest displays of alpine flowers it is best to take the first half of one's summer holiday early in the season, ranging from June till the end of July; but, as one goes higher, the flowering season comes later and we have found many interesting flowers as well as many of the commoner ones during August school holidays above Zermatt and Wengen, both excellent centres. In September the colchicums and other autumn flowers will sprinkle the meadows with colour while many of the main plants will have ripened seed which is probably the best form in which to collect them if one wants a stock. Sow when fresh and allow to freeze during winter in an outside frame where the lights are slightly lifted to allow air throughout the year. Thus one should get many healthy young plants.

I have not yet visited the Alps in autumn but still hope to do so one year. There should be fewer visitors then but, judging from the pictures, the autumn colours, especially of the larches, should be superb. One never ceases to wonder at the prodigality of the flowers which cover the ground and the rocks in a carpet of brilliant colours and seem to live and bloom more profusely for their environment. No man-made rock garden has yet equalled the beauty of a real alpine meadow.

The adaptions of the plants to their conditions of life are more extreme than in all other ranges of plants except perhaps those of the desert and this gives them extra interest. The circumstances with which they have to contend are the strong wind against which no tree could stand and the very short growing and flowering season during which the ground is free from snow. During the long winter they rest dormant below the snow and little harm comes to them. The buds are ready formed before the snow melts and then in a few days from the melting of the snow they come into flower as the warmth unfreezes the ground and the melting snow provides moisture. Often the flowers come up through a hole in the snow, melted possibly by the warmth of their growth. This is especially true of the crocuses in the very early spring and the delicate little soldanellas. Many alpine plants have been found to possess an extra strong concentration of sugars in their cell sap and this assists them both to survive the low temperatures of winter and then to exert stronger osmotic tension through which water is drawn into the cells. Perhaps this is their most important physiological adaptation. Their anatomic adaptations are just as important. Their plant body is reduced to a minimum, sometimes no more than a tussock cushion clinging to the sides of the rocks while the root is very long and tough, creeping between the crevices of the rock, sometimes even enlarging and cracking the rocks as the name saxifrage bears witness: saxifrage in Latin means rock-breaker. Often a plant under an inch in height will have a root system several feet in length. The minimum of resistance is presented to the wind by the plant body while the maximum of anchorage is given by the root.

The wind is used by the plants in their distribution of seed. A very large proportion of alpine plants have seeds that are adapted for distribution by wind. They may have long feather-like appendages like the *Pulsatilla*, or little miniature parachutes like the dandelion and other composites and these may travel long distances. A very large number of seeds are produced in one seed head so that a few at any rate may find suitable resting places, as opposed to the many which may settle on the bare rock or other places where they cannot germinate. Every species of alpine plant is adapted to its own particular little micro-habitat and plant association and it is very rare indeed to find a plant growing out of its environment, be it meadow or scree or rock crevice.

16

SOME FLOWERS OF THE ALPS

(*Top*) *Aconitum napellus*, the common Monkshood, has deep violet-blue flowers which look almost black when outlined against the peaks.

(*Centre*) *Campanula cochlearifolia*, still most often known as *C. pusilla*, is common, but one of the most beautiful and dainty of alpine flowers. The flowers are pale blue, often with a tinge of violet.

(*Bottom*) *Androsace alpina*, often known as *A. glacialis*, grows among the higher rocks and forms cushions or mats covered with little flowers flushed with pink.

SOME MORE ALPINE FLOWERS

(*Above left*) *Fritillaria involucrata* grows in pastures in the Maritime Alps and has large jade-green bells heavily chequered with purple.

(*Above right*) *Phyteuma comosum* has purple flowers each shaped like a wine flask and grows in vertical clefts of the rocks above Lake Garda and in the Dolomites.

(*Below left*) *Campanula cenisia* has slaty-blue flowers and grows in high rocky crevices.

(*Below right*) *Fritillaria tubiformis* var. *moggridgei* grows in the Maritime Alps in damp meadows and has large yellow bells chequered with purple.

TWO FLOWERS OF CORSICA

(*Above*) The white form of *Cyclamen repandum*, a rarity from a valley in the Central mountains.

(*Below*) *Helleborus corsicus* makes large clumps with yellowish green flowers and is common in many parts of the island.

FLOWERS OF THE EASTERN MEDITERRANEAN

(*Above left*) *Cyclamen persicum* makes great clumps in the limestone rocks of the mountains of Lebanon and is abundant, flowering in early spring.

(*Above right*) *Campanula ephesia* grows only on the ruins of Ephesus and on a few neighbouring sites and has large china blue flowers.

(*Below*) *Trigonella balansae* covers large area of the rocks by the sea on Delos. The flowers are deep yellow.

The intensity of colouring among alpine plants is noticeable to every visitor. There is no blue so shattering in its brightness and suddenness, so indescribable in its brilliance as that of the little vernal gentian. It is stronger than any man-made colour, while the blue of the little eritrichium, the King of the Alps, is the clearest imaginable.

My earliest alpine trips when still a schoolboy and university student were to the French Savoy Alps, the country around Pralognan and Moutiers. I travelled out by cross-channel boat and train, the first time with two elderly ladies, friends of the family, who had been for many seasons and knew many of the plants. For most of the time they would sit and sketch the mountain scenery while I wandered off on my own after flowers. Although solitary climbing is not reckoned a wise or good practice, to wander by oneself in the lower alpine meadows and among the rocks up to the snows certainly gives one a savour of the great solitude of the mountains, an impression that has never left me and I have never minded being alone on a mountain. These were good areas for flowers. At Pralognan and again at Mont Cenis, which incidentally is not a mountain but a col over which a road runs beside a large lake, the alpine anemones, now more often placed botanically in a separate genus *Pulsatilla* with the pasque flowers, were abundant and wholly glorious. Here there were the white forms, whilst in other areas one found the sulphur yellow, but rarely the two together. It is not, however, a correct assumption that the white is found only in limestone areas and the yellow where there are granitic rocks, although this more often applies than otherwise. The white form is now called subsp. *alpina* and the yellow subsp. *apiifolia*, although the much better known name of *sulphurea* is still mainly used both by gardeners and by alpine walkers. The flower stands on a stout stem and is a beautifully formed saucer in shape, two or three inches across, with a large boss of yellow stamens in the centre, while the underside is variably flushed with blue; sometimes it is almost covered with blue, at other times it is practically white or just flushed a blue-purple. Below is a little ruff of finely cut leaflets while the main leaves are very finely divided and fernlike. This anemone has a second season of excitement closely following on the flowering. The flower stalks lengthen and the segments of the flower shrivel away but the seeds develop long feathery tails which give the head a tousled struwwelpeter look as well as being an efficient method of distribution.

Just as lovely though much shorter is the spring anemone *Pulsatilla vernalis*. This grows higher up than the alpine anemone, generally in a gritty soil in short grass. The flower is like a pale bluey-mauve opalescent globe from outside, the size almost of a hen's egg, rising out of a furry silvery cup, and opens white to look rather like a tempting poached egg. It is also easier to collect young seedlings, since they have a fibrous root system in place of the long woody rootstock of *P. alpina*. Without making a great excavation it is rarely possibly to get to the base of this where the younger roots emerge. An old chunk will, however, sometimes make more fibrous roots and grow away in a sand frame if treated like a cutting. It can, though, easily be grown from seed. *Pulsatilla* as a genus is separated from *Anemone* by the long feathery tails on the seeds which are derived from a lengthening of the styles after flowering, and by the stem leaves which are fairly close to the flower. In *Anemone* the seeds are a cluster of achenes without any wind dispersal mechanism. This seems quite a good difference and is now accepted by most botanists, although many of us still think of them all as anemones and the garden treatment is similar.

Between the road and the lake at Mont Cenis is a most curious limestone formation where the rock has been worn away to form large holes and the sides are covered with mountain avens, *Dryas octopetala*, which has myriads of white anemone-like flowers with bright yellow centres. It is a shrubby little plant which hugs the rocks and, abundant though it is, one of my favourite alpines. The dark green rugulose foliage complements well the white flowers which seem to be perpetually moving in the wind. With them grow dwarf willows, globularias like little mauve powder-puffs on short stiff stems, and masses of violas both the perky little yellow *V. biflora* and the larger alpine viola, *V. calcarata*, resembling small pansies and varying infinitely in their colour and markings. Around the lake the pink-striped *Ononis cenisia* grows, a dwarf rest-harrow spreading among the short grass. It is a dominant plant over much of the area, but in more moist places is preceded by the pink bird's-eye primula *P. farinosa*, the most abundant of alpine primulas and one of the most beautiful in its clear colour. Later, masses of yellow rock roses flower among the pink ononis.

But even some very rare plants grow quite plentifully behind the hotels at Mont Cenis. One goes up quite a long trek to another little clear lake, frozen for much of the year and tucked into a fold of the land below a great cliff-like cirque of rock. Around it are two lovely campanulas, *C. allionii* with large upright flowers like a miniature Canterbury bell, and *C. cenisia* with smaller slaty-blue bells which almost merge in colour into the shale and slate of the rock and scree. The latter is one of the daintiest campanulas, but needs practically scree conditions for its growth and is rarely seen growing successfully in rock gardens in this country. I have seen it again growing quite plentifully in Switzerland above Zermatt on the way up to the lowest Matterhorn hut, where it had spread into large clumps. Both this and *C. allionii* are usually found on non-limestone shales and continue to flower in the higher regions up till the middle of August. With these is a little spreading viola, *V. cenisia*, with the liveliest expression that I have seen on any plant. I have only found it at Mont Cenis, although it has been recorded at other places in the western Alps; there is a form in the Pyrenees and Maritime Alps, with another in the Apennines. It is usually pale lilac in colour with darker purple markings and the flowers stand stiffly up on short stems, sometimes almost camouflaged among the grey stones.

Another beautiful flower of these higher rocks is *Ranunculus glacialis*, nearly always growing in damp places with snow water running through. It is perhaps the finest ranunculus of all, rather far distant from the ordinary buttercup, The foliage is almost succulent with finely divided leaflets; and sometimes practically sessile, at other times on a short stalk, the white or pink-flushed flowers open to show their rich golden centres. There is something luscious about these thick-petalled, slightly waxy flowers in these inhospitable areas. They are infinitely variable in colour, from white to forms flushed heavily with deep purplish-crimson on the outside; but, like the eritrichium, it is difficult to establish successfully in English gardens. This is not surprising since it has the record for high-living of any flowering plant in the Alps, having been recorded near the top of the Finsteraarhorn in the Bernese Oberland which is 4,274 metres (14,000 feet) high. Other delights of this high scree and shale area are the brilliant yellow *Douglasia vitaliana*, again near-sessile on the ground, and the little purple *Petrocallis pyrenaica*, a small pinkish-purple crucifer which is so much better in the wild than in cultivation, as is also the rather larger and also very variable *Thlaspi rotundifolium*, also found in high areas.

When I first went to Mont Cenis between the wars it was a frontier area. One had to cross the Franco-Italian border before reaching the main lake and there were hazards from mountain troops holding exercises and firing across the higher rocks behind the hotel which sometimes stopped one going to the little clear lake above. Now, since the last war, the border has been moved back further into Italy and there are no military difficulties for the high alpine flower lover to overcome.

In another part one can find cushions of the little blue tufted forget-me-not, with the brightest sky-blue flowers nestling in silvery-grey hairy foliage. This flower is always found high up and I still remember my long walk upwards, lost in a mist and almost ready to turn back, when I suddenly came on whole boulders covered with eritichium in every crevice. The boulders were loose and crumbling and I have never seen Farrer's King of the Alps so plentiful; the flowers I described then as 'the very quintessence of the bluest of all skies, the sky of a summer evening when the sun is fast setting'.

For Mont Cenis one can either stay at the col where there is the old Hôtel de la Poste, extolled in so much of the older alpine-plant literature, or the newer hotel, but these are noisy during the day with many coach parties stopping on their way to Susa in Italy, If one has a car, however, one can stay lower in the valley at Lanslebourg and come up each day, for one will surely want several days in the area. By train one travels to Modane and then takes a Post Coach to Lanslebourg and Mont Cenis.

Another very good area in the Savoy Alps is around La Grâve near the foot of the great Meije, and Lautaret which lies on another pass between Grenoble and Briançon. One can approach Lautaret from Modane and Lanslebourg by going over the Galibier Pass and here in late June and early July are thousands of the little white *Ranunculus pyrenaeus* carpeting the ground like snow, continually blown by the wind. They are all white but some are small and single in flower while others are semi-double and like little paeonies though only an inch across and with stems three to six inches tall. They come into flower soon after the snow melts; there can, however, be a lot of snow on the Galibier Pass in June and it is in some seasons not clear until July. The same applies to the masses of alpine crocus, *C. vernus*, so much more delicate than its garden hybrid descendants, but alas no easy or satisfactory plant for our garden. If we want wild species crocuses in our garden we should grow masses of the very early spring *Crocus tomasinianus* or the autumn *C. speciosus*.

At Lautaret there is a famous alpine hotel but it is on rather a bleak open col and, provided one has a car, one may prefer to stay a little lower down at La Grâve with its incomparable view of Le Meije and in itself a good centre for walks on the slopes of that mountain. In the woods above La Grâve the lily of the valley has spread widely. At Lautaret there is a good alpine botanic garden maintained by the University of Grenoble and containing a great variety of alpine plants. It is well worth a visit. My other chief memories of Lautaret are the eritrichiums up one of the side valleys here growing in a more cliff-like formation. There are also great thickets of *Daphne striata*, reddish purple and waxy, sweet scented, but also very difficult to grow successfully in England. Another speciality of Lautaret is the fine tussocks of *Dianthus neglectus*, that beautiful dwarf pink with the fringed edge to the flower.

Another good area is around the Val d'Isère and the Col de l'Iseran where the road goes up over 9,000 feet and is one of the highest roads in western Europe. I stayed several times at a delightful little hotel by the Lac de Tignes looking out to the magnificent snow ridge and dome of the

Grande Motte, but I have not been back recently and I understand that there have been extensive hydro-electric works in the valley. One could stay then quite comfortably at the little hotel for between five and ten shillings a night, but that was in days before the last war. One of my pleasantest recollections is the delightful walk over from the Val d'Isère to the Lac de Tignes through woodlands full of splendid forms of *Atragine alpina,* the beautiful mountain clematis with its pendulous mauve-fringed bells.

Many climbers pass as quickly as possible through the woodland zones, but to me these offer some of the most attractive parts of the mountains and the perfect contrast to a day among the eternal snows. The hardwood trees, beeches and oaks and rowans, seldom reach above 4,000 feet and this is usually described as the sub-alpine zone; but in the early spring this zone contains some of the most beautiful of plants, particularly in the glades where a clearing may bring extra light. In the very early spring there will be hellebores, varying forms of the great white Christmas Rose, *H. niger,* but often tinted with delightful shades of pink and maroon on the outside of the petals. Here also grow the hepaticas, little anemone-like plants with trilobed leaves, predominantly sky-blue but again varying from white to deeper shades of blue and mauve with even an occasional pink. It is worth selecting the best forms when they are in flower. We were lucky enough to find a great abundance of these and the hellebores growing and flowering in early April practically down to the lake level among the hills around the lake of Como. Higher up they will flower rather later. In the same area we also found the round-leafed *Cyclamen europaeum,* at this spring season betrayed by its leaves carrying a promise of rosy-pink flowers to come in plenty in the autumn. It is the finest scented of all the cyclamen and a valuable species as a garden plant, although curiously requiring rather a more protected position than the more southerly Italian *C. neapolitanum* which grows around Naples and southwards. In cold gardens the corms should be tucked in under a low-growing shrub or a rock.

Later these hills will be parched and brown, but as we move upwards into the mountains proper there seems to be ample moisture and luscious greenness everywhere: no meadows are so bright and full of colour as the alpine ones. The terrestrial orchids feature among the glories of the wood-land and lower meadow zones. Among beechwood primarily, and the chalk, the very lucky traveller may happen on Venus's Slipper, the nodding, wide-winged flowers of *Cypripedium calceolus* with their ochreous-yellow pouch and brown-winged petals. This is a rarity indeed and if found should be photographed with care but not collected. In Switzerland it is very rightly protected by law and should be spared in other countries. If you want to establish a clump in your garden – and it is not very easily established – it is better to buy a plant in a pot from one of the English alpine nurserymen who specialize in such plants.

The spotted purple orchids are largely plants of the damper meadow and in some places are abundant as well as being very variable. Again, they do not transplant well but this can sometimes be managed if care is taken not to damage the young tuber growing from the old one. This is the part from which next year's flower and leaves will come. In Switzerland most are rightly protected. Among my favourite alpine orchids are the little round deep crimson heads of the vanilla orchid, *Nigritella nigra*, and the sweet-scented pale lilac mauve spikes of the *Gymnadenia odoratissima* and *G. conopsea*. These are, however, dwellers in the higher alpine pastures where often they are plentiful and they seem to be spread like the alpine anemones all over the Alps.

The woodland zone and glade are also the home of the martagon lily – great spikes of nodding Turk's-cap purple lilies, sometimes from old bulbs growing five feet in height and bearing a very great number of flowers. In the Savoy, and also in the Maritime Alps, one will find the upturned and brilliant fiery orange cup-shaped flowers of *Lilium bulbiferum croceum*. These latter are rather temperamental but the martagon are good growers in English gardens. Rather than collected they should be bought from a bulb merchant or raised from seed.

The wild clematis of the Alps, *Atragine alpina*, with its beautiful nodding soft grey-mauve flowers is often found rambling over the rocks and through the undergrowth. The flowers are small in comparison with the soup plate clematis of gardens, also beautiful in their own way, but they have delicacy and grace which is very pleasing. We found them again on the Mont Cenis pass, scrambling over the very curious limestone pot-holes which abound near the edge of the lake there. This is indeed a rich region for alpine flowers.

The alpine meadow, which is scythed with such care and skill in the summer, gives us a picture in early June that is unrivalled for colour by any meadow that one may see in England. Here nature has achieved such a skilful blending of colours as is found in few herbaceous borders. The deep blues and mauves and even magentas of the variable *Geranium pratense* contrast with the yellow trollius, if the meadow be a damp one, and sometimes also with masses of dwarf poeticus white narcissus. These are without the very strong sweet scent of the tazetta narcissus of the Mediterranean and north African meadows but still not altogether lacking in that quality.

I remember very well one particular meadow which was thick with white poeticus narcissus and purple spotted orchids, a delightful combination which I can strongly recommend to anyone having a small damp meadow which is, perhaps, cut for hay but not grazed. It would only be likely to be successful in damp ground and it is possible that in this country the narcissus might sometimes flower before the orchid. The golden trollius would also grow well under the same conditions.

In drier places – but remember few meadows in the Alps are really dry for long with the water draining down from the melting of the snows – masses of the mauve *Salvia pratensis* often grow, and with it one can sometimes find the delicate white St Bruno's lily, *Paradisia liliastrum,* or the even smaller St. Bernard's lily, *Anthericum liliago,* both like a miniature and much frailer Madonna lily. Here also and at the edges of the forests or in sunlit glades are to be found the great pleated leaves of the veratrum, *V. viride,* and the big yellow gentians, *G. lutea* and *G. punctata,* in both cases much more decorative in leaf than in flower. For this quality the veratrum is well worth growing in English gardens although one rarely sees it. In the spring it comes through the soil with an exciting vigour like an *Eremurus.*

On the damp rock faces are often to be found choice primulas and sometimes androsaces, the deep mauve purples of *P. viscosa* and the *P. pedemontana,* a local plant of great beauty with clusters of pinkish-purple flowers, each with a white eye and a little russet furry edge to the leaf which makes it easy to recognise. It may be found around Mont Cenis. Here too will be the hanging mats of *Saxifraga oppositifolia,* deep glowing stars of rich purple, shining against the dark rock. It is also sometimes a plant of screes, where it makes roots often to be measured by yards rather than by feet. Another plant distributed likewise all over the Alps, and generally to be found in the same environment of loose scree, is the alpine toadflax, *Linaria alpina,* with masses of little mauve and orange flowers well set off by the blue-grey foliage. It does not transplant well, and is rarely seen in English gardens. It may, however, be treated as an annual, the seed being sown directly into a scree.

Generally one does not find the *Androsaces* below about 7,000 feet, although the white *A. lactea* may occur lower in short turf. The most exciting, however, are the cushion and tufted dwarfs with almost sessile flowers. *A. alpina* has flowers of a delicate shell-pink and grows in damp shales. I have seen it in the Savoy Alps and it is plentiful along the ridge which runs below the Kornergrat station above Zermatt in the direction of the Matterhorn; it is one of my favourites. Other more dense compact cushions such as *A. helvetica* or *A. pyrenaica* are inhabitants of damp north-facing and vertical crevices, where they rarely get full sun and where there is much cloud. The flowers are like little white stars with yellow eye-like centres. They are favourites of alpine house gardeners, but only the smallest seedlings are likely to transplant successfully and in cultivation they must never be watered overhead. Some others are much more local, while a few naturally occurring hybrids have been noticed. The rather looser growing pink *A. carnea* and its forms are generally easier to manage in cultivation and the type is quite widely distributed in both Alps and Pyrenees, where they are inhabitants of rather firm scree and scree-like very short grass. With them often grow some of the most delicate of the alpine ferns such as the Finger Fern *Ceterach officinale*, each leaflet rimmed with silver. These are especially plants of the rock crevices.

Another lovely group that one finds among the mountains is the soldanellas, the delicate bell-like flowers with the fringed edges that come up so plentifully as the snow melts and constantly seem to blow in the wind. The commonest ones in the central Alps are *S. montana* and *S. alpina*. Sometimes one will find them actually growing through a patch of snow, the very slight extra warmth which their growing gives out having melted the snow. Like the androsaces they belong to the *Primula* family and sometimes they are known as 'Snowbells'. *S. montana* is the larger with leaves described as up to three inches across, and flowers violet-blue or lilac-blue, or sometimes a slaty white, three to six to a stem, the bells being fringed almost for their complete length while the bells of *S. alpina* are slightly smaller and only fringed for half their length. *S. alpina* is found more generally in the western Alps – while *S. montana* has a more easterly distribution. Even smaller and more delicate are *S. pusilla* and *S. minima* but I have only found them in the Dolomites and eastern Alps and so leave their description to the section on that area.

Gentians and saxifrages are two of the great genera of the Alps and they are both numerous in total numbers as well as in variety. The plant known as *Gentiana acaulis* is infinitely variable, and if one accepts the differentiation into several species perhaps the name *acaulis* is better given up although it is well rooted in gardening literature. In most modern floras of the Alps it no longer finds a place, being superseded by *G. clusii*, the deep sky-blue trumpet gentian from limestone areas; and *G. kochiana*, usually from more acid soils and generally with deeper blue flowers, rarely of such a fine colour as *G. clusii*. These are probably the commonest gentians in the Alps, but the sight of them never palls with their brilliant blue upright trumpets. What a subtle, almost unreal colour to associate with a flower, yet there it is. Even brighter blue is the *Gentiana verna*, especially in its higher and more dwarf forms *G. brachyphylla* and *G.*

bavarica. These make thick clumps of closely clustered rosettes, sometimes absolutely solid with flowers. The flowers are generally a less deep blue than those of *G. verna* although even more brilliant. *G. brachyphylla* is also distinguished from *G. verna* by the absence of any wings to the angles of the calyx which is much more slender. Even more jewel-like is the little annual snow gentian, *G. nivalis.* The flowers are minute, only ten to fifteen millimetres across, but they gleam like jewels. It is so small, though, that it is difficult to photograph adequately. The willow gentian, *G. asclepiadia,* is a tall plant of the sub-alpine woodlands and the flowers lack the shattering bright blue of the higher gentians; nevertheless, when naturalized it is a pleasant enough plant for the woodland garden. I always find interesting the statuesque gentians of the *lutea* group. These seem so remote from the dwarf blue species. *G. lutea* is the finest, with spikes sometimes nearly up to six feet, covered for all the upper part with bright yellow flowers while the basal leaves are large and pleated like those of a veratrum. It is surprising how rarely one sees this gentian in English gardens since it is a good perennial. From its roots the liqueur associated with gentians is made. Purplish or coppery-red is the smaller species *G. purpurea,* and this is particularly fine on the meadows above the Kleine Scheidegg below the might of the Jungfrau. I remember its association with the big thistle *Cirsium spinosissimus* in which every spine gleamed like silver when seen with the low sun behind it. It is another plant of great distinction but one best left in its native habitat.

The prostrate thistles with the great central flower heads, which are so decorative when dried and in seed, belong to another genus, *Carlina,* and are *C. acaulis.* When in seed the stiff bracts around the centre gleam like burnished pewter, hard and metallic. As a garden plant it is sometimes offered but it is a biennial only. Other plants which look best when seen against the light are the viviparous grass, a form of our ordinary *Poa annua,* the commonest of grasses, in which the seeds germinate when still in the seed head and begin to grow, giving the young plants a start in distribution; and the bog cotton, spectacular in the Alps as it is on Scottish moorlands. The very dark blue monkshood also seems most decorative when silhouetted against a low sunlight with the snow behind.

The second of the two greatest genera in these areas are, as I have said, saxifrages. They are easily divided into three main groups, although there are others which are not so colourful and are not often collected for garden plants. The most typical of mountain saxifrages are the encrusted or silver saxifrages which include such fine plants as *S. aizoon,* variable and abundant nearly everywhere as a rock plant with its small plumes of white or creamy flowers, sometimes very finely spotted with red, and neat encrusted grey rosettes. The little encrusted spots which edge the leaves actually consist of little bits of limestone. An experiment showed that they could be dissolved in a foaming, fizzing bubble by a drop of acid. Perhaps this is nature's way of getting rid of material which the plant does not need, although making it more beautiful in our eyes during the process. Much larger and much rarer are the great rosettes and vast plumes of *S. longifolia* of the Pyrenees and *S. lingulata* of the Maritimes. I do not think that I have ever found the latter, but the Pyrenean plant will be mentioned in the chapter on that range. All these are fine garden plants especially for growing in the crevices of a stone wall. Sometimes one meets great clumps of rosettes of *S. aizoon* on the rocks with their plumes all blowing in the wind and very lovely they are. The *Porphyrion* group contains the well known *S. oppositifolia* which often cascades down north facing and damp crevices among the rocks, the innumerable cup-shaped flowers gleaming like purple or red jewels. Even more jewel-like is the little *S. retusa* which usually has more red in the purple of its flowers. Both species, however, are very variable. I particularly remember some very fine reddish forms of *S. oppositifolia* spreading flat over the rocks behind the Mont Cenis hotels, a rather unusual method of growth for this species; still, the roots were well down in the crevices. These are also plants of the damp screes, but never on dry places. In cultivation there is also a delicate and lovely white form but I have never seen this in the wild.

The other abundant saxifrage is *S. aizoides*, the yellow mountain saxifrage, and this is placed on its own in the section *Xanthizoon*. It also grows in Britain and in the west of Ireland, nearly always in damp places where the water is moving, such as moist shingles of stream sides or on the banks of streams. It grows into large masses, sometimes a yard across, covered with bright yellow starry flowers each one faintly dotted with red, and has a long flowering season from June to September.

There are so many other flowers of the mountains to mention: the little yellow cushion-like drabas, the silvery white flannel-like stars of the edelweiss, now regretfully becoming much rarer but by no means confined only to inaccessible and precipitous places. Then there are all the mauve and purple calaminthas and scutellarias which grow especially on the screes, the bright yellow *Douglasia* which forms cushions of yellow at the edges of the screes, the ubiquitous alpenrose *Rhododendron ferrugineum* and *R. hirsutum*. An alpine holiday would not seem complete if we had not seen them. Usually, but not invariably, *R. ferrugineum* is confined to peaty areas among acid rocks and *R. hirsutum* to the limestone areas. It is easily distinguished by the hairy leaves, green below. Both species are equally variable and fine in flower, sometimes covering large areas with a low scrub through which one can nearly always plunge. The finest stands I remember were above Saas Fee in Switzerland in a little side valley at the top of which were eritrichium and aretian androsace, a long but worthwhile trek. Unfortunately the eritrichium rarely had the sun off it and my photographs showed only a pale and disappointing washed out blue. Generally these very brilliant blue flowers are best photographed away from bright sunshine. A close relation of the alpenrose is the little alpine azalea, *Loiseleuria procumbens*, creeping prostrate over the rocks and covered with minute pink stars.

The alpine poppies were very fine and delicate in some of the Savoy Alps, and varied in colour, but even larger were the yellow forms of the Eastern Alps and Dolomites, usually grouped under *Papaver rhaeticum*. Always, however, the colours seem to be clear: white or cream, pale yellow, deep yellow, pale pink and even an orange-red, as variable as the larger Iceland poppies. The tussocks of little grey leaves, so finely divided, flower abundantly on slender stems swaying in the wind. There are few alpines annually so beautiful.

The moss campion *Silene acaulis* probably would earn the prize for the largest number of alpine flowers in a small area. These bright pink stars jostle each other so tightly over the cushion that all the green is hidden – a miracle of floriferousness, but perhaps not so pleasing artistically as other plants which do not flower so abundantly. Sometimes one finds pink cushions a foot or more across. It is curious that a plant so free flowering and so widespread over a large range of altitude in the Alps rarely flowers satisfactorily in lower gardens.

Another plant that one rarely misses is the beautiful little campanula *C. cochleariifolia*, which used to be known as *C. pusilla* or in some books 'Fairy's Thimble,'. This describes well the little blue bells dancing on their delicate stems. It is primarily a plant of damp screes but it is common in all sorts of piles of loose grit by the path side or creeping in the cracks of rocks. It is very variable in colour, ranging from quite clear blue to a violet-mauve, and occasionally white, but is always delicate and delightful. Campanulas also form a main constituent of the alpine meadows, *C. rotundifolia* the harebell, *C. linifolia* with the narrow leaves, and *C. scheuchzeri* with narrow leaves and more violet-purple in its broad bells. They give meadows much of the blue and violet colouring which makes them so brilliant. Their charm, however, lies in the variety of colours rather than in a mass of any one colour. Completely different among the campanulas are the straw-yellow spikes of *C. thyrsoides*, stout and massive, statuesque rather than graceful, but very distinctive when they form large groups, each spike a foot or two tall. The flowers are almost stemless, clustered thickly round the flower stalks. Another favourite of ours is the bearded campanula, *C. barbata*, with its delicate powder-blue bells on stout hairy stems; they are easily distinguished by the fringe of hairs round the lobes of the bell and are fairly common in most areas. Occasionally one finds a white one which stands out clearly among the others but I have never seen a colony all of this colour.

THE MARITIMES

This is the area behind the French and Italian Riviera towns and it is very rich in interesting plants, the snows melting much earlier here than in the Savoy and Swiss Alps. My experience of it is not so great but nevertheless some plants still stand out in my memory. On a drive into the mountains from Menton in early May, guided by Miss May Campbell, we found both the fritillaries which are endemic to the area, and in fairly large groups. These are *F. involucrata* and *F. moggridgei*. The former grew below 4,000 feet and had large single bells very variable in length and colouring even in the same patch. Their basic colour was a jade green and they were chequered with reddish-purple. One, which I photographed, was almost black with its chequering. The stems are delicate, six inches to a foot high, and the flowers, which are usually single, have three long leafy bracts sticking out above them at right angles to the stem. They are not quite so large as in our native Snake's Head fritillary but equally lovely and delicate. It is recorded as growing also in Provence and northern Italy but I have not seen it there. *F. moggridgei* is now more often regarded as a variety of *F. tubiformis* and grew in higher areas than *F. involucrata*. We found it plentiful in one valley of short grass where its pale yellow bells nodded on short stems. The flowers are almost as wide as long and are chequered outside with dull purplish-red. It is a lovely plant which is very local in its occurrence and northwards is replaced by the purple-flowered type of *F. tubiformis*. Farrer reported one area where they grew together but I have not seen this.

We found also masses of dog-tooth erythroniums in flower, quite good pinkish-purple forms with well marbled foliage. What a great deal the quality of the foliage adds to this flower! With them were the first gentians of spring, and lower down a fine spurge which was apparently a form of *Euphorbia hyberna*, jade green and yellow in its bracts. This also grows in Spain and the west of Ireland, but is rarely seen in cultivation.

On another visit we approached the rather dank cliffs on which grew *Primula allionii*, so great a favourite of alpine house owners. The plants grew literally in cracks in the cliff and mostly were quite inaccessible; however, none of those we saw were nearly the equal of some of the fine selected forms in cultivation nor did the pinkish-purple flowers cover the tussocks completely as in some cultivated plants; but still, it is always a thrill to see such a plant growing in its wild habitat, and well worth the rather long trek which we took that day. Another memory of these sun-baked mountains is the brilliant hot orange, almost a vermilion-red, of the upright chalices of *Lilium bulbiferum*, never very tall and usually single in flower, growing on steep slopes which were covered in low prickly scrub. These were without bulbils and different from the taller more orange and less red form of farther north which is var. *croceum*. Another distinction was that the segments of the flowers were more claw-shaped at the base, showing light between. Nearly always they grew singly, never in clumps or clusters, and I have not seen this form in cultivation although it must surely have been tried.

The Maritimes also have most of the commoner alpine plants, *Dryas* and *Soldanella, Gentiana verna, Saxifraga aizoon* and *aizoides.* The brilliant scarlet lily, *L. pomponium,* still grows there, although I have never seen it. A friend who knows the area well once sent a few bulbs and said that they were plentiful in the valley where he collected them. I do not, however, have its exact location, so cannot reveal it. This lily is easily distinguished by the crowded narrow leaves and the shining Turk's-cap flowers of the brightest red. It rarely grows more than three feet tall, but in general now it is a plant which should be left *in situ* and grown from seed.

Saxifraga lingulata is undoubtedly one of the finest of the great plumed encrusted saxifrages. Farrer, in his superb book *On High Hills,* gives a good description of it which I cannot rival, especially as I have only seen it in cultivation and not in the wild. It is a plant of the limestone rocks and cliffs. In general the Maritimes form is the variety *lantoscana.* It is a variable plant though, and other forms, sub-species and varieties occur in the Cottian Alps farther north. The rosettes are large and loose with narrow leaves. By the looseness of the rosette it is easily distinguished from *S. longifolia* of the Pyrenees which has even larger plumes. All this group have white flowers on pinkish-crimson stems which branch freely. The rosettes are twice the size of those of *S. aizoon.* The main rosettes die after flowering, but usually leave behind numerous little side rosettes which take several years to grow to flowering size.

Another interesting plant of this area is *Primula marginata,* easily distinguished by the mealy white edges to the leaves and the pale violet or lavender flowers. It makes a hardy almost woody rhizome and is a plant of rock crevices and cliffs up to 6,000 feet. From this, good garden forms have been selected among which 'Linda Pope' is an old favourite.

The lower zones are colourful with interesting genistas and helianthemums and the little blue *Aphyllanthes monspeliensis* which has rush-like stems and a few leaves of the same type, while each stem has a single blue starry flower with six spreading petals. It grows around the western Mediterranean coast hills and in northern Africa. I only remember noticing it particularly in the Maritimes, but it is very inconspicuous when not in flower. It belongs to the lily family. Lower down to the level of the Upper Corniche road in the Nice area, but now becoming very rare owing to the great mass of building all along the coast, is the beautiful little white *Leucojum nicaeënse*, a snowflake only three or four inches high with open white bells and narrow glaucous leaves. It is distinguished from all other species of *Leucojum*, except the excessively rare *L. valentinum*, which I have never seen, by a curious six-lobed disc at the base of the flower. It is a tender plant in most English gardens but grows well in alpine house or bulb frame and should be looked after and propagated from seed since it is now becoming so rare in its native habitat. It used to be better known under the name *L. hiemale*, which completed the seasons neatly for the genus, and it is a pity that the more awkward name *L. nicaeënse* has had to be adopted, but it is nevertheless a plant of natural grace and well worth growing. It is not always, however, very long lived in cultivation.

THE DOLOMITES AND EASTERN ALPS

My memories of the Dolomites return to spectacular craggy peaks, much mist and rain and, among the flowers, particularly to the very bright pink clumps of *Primula minima* and some lovely forms of *Primula auricula* growing on damp cliffs among the vertical sides of the old army trenches and fortifications on Monte Piano. That delicate little relation of the rhododendrons, *Rhodothamnus chamaecistus*, was abundant, and among the screes *Soldanella pusilla* and *S. minima* were finer than I had seen anywhere else. So were the larger yellow forms of *Papaver alpinum* which we usually call *P. rhaeticum*.

We stayed at Misurina by the lake as so many alpine gardeners have done, and climbed up a rough track to the foot of the Drei Zinnern. I believe that Misurina has been much developed with more and even larger hotels and probably much of the rough track has now become a road for cars. But still, these truly vertical pillars rise sheer from the scree ridge and at close quarters like this are some of the most impressive peaks I have seen. They are certainly just like three gigantic fingers or, as Farrer described them, like flames. The colouring is grey and pink with large patches or streaks of black, caused presumably by oxidation of some of the chemical in the rock: at any rate it is most effective. When one gets up to the base of the rock one sees that it is not quite so smooth and sheer as it appears from a distance and is in places honeycombed by possible hand-holds. The little Zinn is reputed as the most difficult climb while the Great Zinn is the taller. The ridge below them is made up of limestone detritus and presumably has perfect drainage. In it are the finest specimens of *Primula minima* I have ever seen. The flowers are quite large, like those of *P. allionii* and a good bright pink, easily distinguished by the deep clefts between the petals, while each petal is divided into two distinct lobes. The leaves also are lobed and fringed. We saw it again growing plentifully in the short grass beside the higher stretches of the Gross Glockner pass, one of the finest roads in Europe for seeing mountain flowers. It is not, however, an easy plant in cultivation and good ones are rarely seen. Some have found a white form but we were not so lucky. The wild auricula is mainly a plant of the limestone and seems to grow best in clefts in vertical rock faces. It is not a plant of very high altitude. It is certainly one of the most lovely of its richly endowed genus, *Primula*, with its thick green leaves hugging the rocks and covered in a white mealiness, while the soft yellow flowers, each with a white eye, rise from the centres of the rosettes. A good specimen of the wild form is to my eyes as lovely as any of the choice forms which have been raised from it. It is a plant much illustrated in the old herbals.

Here also was the little silver leaved potentilla, *P. nitida*, one of the most beautiful members of the genus, but very variable. The foliage is really silvery in prostrate mats and the flowers are sessile or with short stalks, in the best forms a deep clear pink and about an inch across, in other forms a paler pink. In cultivation it is disinclined to flower freely but again forms are variable in this respect.

The *Rhodothamnus* was a most charming little plant forming a mat of woody branches and small leathery and hairy leaves with erect pink flowers each nearly an inch across, saucer-shaped and almost flat, sometimes several together in a small cluster. It is close to *Rhododendron* and was formerly placed in that genus and known as the dwarf alpenrose, but the petals are separate without any tube as in rhododendrons. It was plentiful but in cultivation it has generally proved rather a difficult plant.

There were *Cyclamen europaeum* in the woods and white *Anemone trifolia* close to *Anemone nemorosa*, but a little larger in flower and without the finely divided leaves. By Misurina one could sometimes find forms flushed with pink or mauve.

In the woodlands were also *Pyrola rotundifolia*, the round-leaved winter-green with quite tall stems of white saucer-shaped flowers, pendulous like more open lily of the valley bells, only twice as large. They were never common but always striking. Unfortunately in cultivation all pyrolas seem difficult, although occasionally they will settle down in almost pure leaf mould. Higher up there the *Pulsatillas* were very fine, both the white and yellow forms of *P. alpina*. Another strong memory also brings back the higher scree with the delicate little white bells of *Soldanella minima* flowering by patches of snow, each flower fringed round the lip. Only a very little larger was *S. pusilla* with purple flowers. There were also the commoner and much larger *S. alpina* and *S. montana*, but it was the little ones only two inches or so tall that excited me most. With them grew masses of the little yellow alpine poppies, some of the most graceful of all plants with their fernlike silvery-grey foliage. The yellow form, *Papaver rhaeticum*, seems to have slightly larger flowers than the more common white and pink forms.

The flowers in the mountains and meadows around the Italian lakes are also of interest. One spring early in April we stayed on Lake Como and saw masses of finely coloured forms of the Christmas Rose, *Helleborus niger*, some a good white, but many flushed with shades of pink and even deep maroon on the outside of the petals. Higher up in the Julian Alps we saw later even larger forms, mostly white and taller and larger than our usual form; they were probably the one known as *macranthus* and very lovely they were, great white open saucers several inches across. It is only in the Eastern Alps that one finds these forms.

One summer late in June and into early July I drove with Eliot Hodgkin to visit the Julian Alps in northern Yugoslavia, stopping on the way for a few days by Lake Garda in Italy and driving up into the mountains to the west of the lake. Surprisingly low, only a little above the lake level, we found that oddest of all flowers *Phyteuma comosum*, which earlier I had failed to find around Misurina. Here it was growing in clefts of vertical limestone, large tufts made up of mauve flask-shaped flowers, deeper in colour at the neck and paler at the globular base. Each flower is like a miniature Italian wine bottle and is almost unique in flower form and is sometimes known as the Devil's Claw. From the end of the flask the stigma protrudes. It is only found in the eastern Alps but usually below about 6,000 feet; previously, I think, we had looked for it at too high an altitude. Rather high on the limestone crags and safely inaccessible we found, but rather sparsely, that most exciting of the daphnes, *D. petraea*, which is restricted to only a small area in these mountains. The flowers were in tight heads of deep pink, waxy and shining, but I do not think that we saw any forms as fine as some of the best in cultivation and certainly no specimens nearly as large. We also looked for another rare endemic of the area *Silene elisabetha*, formerly known as *Melandrium elisabethae*, a perennial campion with large deep pink flowers, but we did not find any good

FLOWERS OF CRETE

(*Above left*) *Tulipa bakeri* on the Plain of Omalo in the White mountains.
(*Above right*) *Ebenus creticus* is a large pink-flowered clover with silvery leaves and grows only on Crete. (*top right*)

(*Below*) *Ranunculus asiaticus*, the white form, is the dominant one in Crete.

FLOWERS OF THE EASTERN MEDITERRANEAN ISLANDS

(*Above left*) *Crocus cyprius* grows around the top of Mt. Troodos in Cyprus.
(*Above right*) *Cyclamen creticum* is abundant still in the woods of Crete.

(*Below left*) *Ophrys sphegodes* is lovely on Rhodes and Crete and is a variable
species.
(*Below right*) *Gladiolus segetum* has bright rosy-purple flowers and is abundant
in some areas of Crete and Cyprus.

specimen in flower for photography. Sometimes the flowers are two inches across but in cultivation it is also very rare. These lower mountains are seldom visited by plant lovers but nevertheless they do contain some good plants.

The Julian Alps are in the north-west corner of Yugoslavia only a little over the border from Austria, and centre round the Triglav massive. We stayed at a cost of only five shillings a night at a little village near Bled, a small town beautifully placed on a great lake. Triglav itself is over 9,000 feet and full of interesting plants. Our most exciting find there, however, was *Lilium carniolicum*. We drove to Lake Bohinj at the foot of the mountain and then slogged up the steep zig-zag, partly wooden steps, partly steep path, of the side of the mountain beside the lake. It was a very hot day and I remember well how gruelling it was. However, where it levelled out a bit at about 5,000 feet we came on a mass of boulder scree, rather an unusual formation, and in this unlikely environment for a lily we found *L. carniolicum*, good spikes up to three and a half feet, each with half a dozen or more bright apricot-red or cinnabar-red flowers, pendulous and Turk's-cap like a martagon with stout waxy petals. They were slightly green at the base with rays flushing out into the centre of the petals. It was a very handsome flower and it was a great thrill to find it. The path brought us to a hut and some delightful little lakes where patches of snow still lay, but auriculas flowered in great masses among the *Rhodothamnus*.

Another good area of approach to the mountains was over the Vrsic pass from Krajanska-Gora, and here the flora was very rich in late June. One can also visit there the Julian Botanic Garden but it did not rival that at Lautaret or Schynige Platte. The alpenrose here is mostly *Rhododendron hirsutum* rather than *R. ferrugineum*. Another plant of the eastern Alps which we found growing abundantly on the lower rocks of Triglav was *Potentilla nitida* which makes mats of beautiful silvery foliage covered with open pink flowers on short inch-long stems. The flowers varied in colour from a pale shell-pink to quite a good deep pink. It is a charming plant but usually rather shy-flowering in cultivation. Another unusual plant of the Triglav scree was the biscuit-yellow dwarf *Alyssum ovirense* and I had not seen this elsewhere. It is named after the Hoch Obir of the Karawanken Alps. It was a delightful plant with silvery-grey foliage spreading mat-like over the rocks, and quite large flowers of an unusual soft shade of buff-yellow, far removed from the strident bright yellow of *A. saxatile* which we tend to associate with the genus. Curiously, here also growing with the rhodothamnos was *Erica carnea*, quite good crimson forms, showing this plant's tolerance of the limestone. Another plant which I did not see elsewhere was *Aquilegia einseleana* with rather small flowers of a deep purplish-mauve and finely divided foliage, not so fine a plant as *A. alpina*, but possibly easier to grow in cultivation. The red helleborine, *Cephalanthera rubra*, also grew at the edge of woodlands and was a fine species since the flowers opened much more than in some of the other helleborines. They are not red, however, but a strong purplish-pink. Since it is so rare a native of Britain that few of us are likely ever to see it here, the continental colonies must suffice for us. The two lateral petals spread out to give us a flower one and a half inches wide but only an inch or so in depth. Another exciting plant of the area and of the neighbouring Karawanken Alps was *Campanula zoysii*, a rare species with narrow flask-shaped blue-mauve bells, generally growing in vertical clefts in the rock, but occasionally washed down by a stream to lower patches of grit by the river side. Unfortunately it is difficult in cultivation.

We drove also for a short distance down the coast to the hot dry karst country behind Trieste. This country consists of limestone hills where the meadows are curiously pitted with dolines – big holes five or six yards across and as deep. Here the most interesting plant for us was the large form of *Gentiana verna* known as *G. tergestina*, a very fine plant but only slightly distinct from the type, which at any rate is infinitely variable. There were also dwarf forms of the most brilliant orange-red *Lilium bulbiferum*.

I have unfortunately no experience of the southern Yugoslavian mountains but they also are known to contain many interesting plants. There are always new areas to visit.

CHAPTER TWO

The Pyrenees, Spain and Corsica

THE PYRENEES

The flowers of the Pyrenees are close to those of the Alps yet there are some interesting endemics which one finds nowhere else, so that a separate visit is well justified. In general they are also very pleasant mountains, not quite so vast in scale as the Alps and with long gentler valleys. Their flora, especially at the eastern end, includes some of the Mediterranean plants, while on the western end and the southern side there are Spanish plants. We went first to the northern side, early in July one year visiting Font Romeu, a small village just over 5,000 feet which is near the eastern end; then we went on to Gavarnie nearer the western end, these being two of the best known centres although there are other good ones. Recent development of Font Romeu as a skiing centre has enlarged the village into a town since our visit. Later I stayed with Eliot Hodgkin and Bertram Anderson at the Parador at Ordesa on the Spanish side, again towards the western end, and drove back across the range by another pass. We also visited Andorra from the northern side.

One ascends to Font Romeu from the French side through delightful little old towns such as Mont Louis which is well known for its dwarf daffodils in April. Near Font Romeu the best known site botanically is the Val d'Eyne, a broad valley which starts just below the village and leads up towards the Spanish frontier by the Col de Nuria. Even as late as July there were still meadows full of *Narcissus poeticus* in flower and very lovely they were, narrow and twisted in petal as compared with the best garden forms but still as strongly scented. The range in altitude and so in the season of flowering of these daffodils is very great. Lower down they would probably have been in flower in April. The particular plant of this valley is *Adonis pyrenaica*, and while plentiful enough in some areas it is not abundant. If collected, only seed or small seedlings should be taken for it is notoriously difficult to establish in cultivation. The foliage is fine and fernlike similar to that of a fennel and the flowers are deep clear yellow about two and a half inches across, many-petalled and rather globular when not open, like a much larger buttercup. It grows among the rocks with stems about a foot and a half tall and is certainly one of the most

35

beautiful species in its genus. Besides the eastern Pyrenees one other station is recorded for it in the Maritime Alps. There seems no valid reason for its reputed difficulty in cultivation for it is not a very high alpine, while plants around it such as the large *Pulsatilla alpina* grow easily enough. The blue columbine *Aquilegia pyrenaica* is also a difficult plant to cultivate but it is staggeringly beautiful among the grey rocks of its native Pyrenees. The flowers are clear blue, rather pale, and sometimes with a slight lilac tinge, and with slender curved spurs at the base. The stems are slender and they nod gently in the breeze.

Another Pyrenan speciality of the eastern end and central part of the range is *Gentiana pyrenaica*, not the bright azure of *Gentiana verna* or even the darker colour of *G. acaulis*, but a deep violet-blue with flowers opening more nearly flat than in these other gentians, and an appearance of ten lobes since the intermediate lobes are nearly as large as the main ones. It is quite distinct and seems to grow freely in peaty boggy places on open moorland. We found it fairly common by the road on the pass into Andorra from the north. In cultivation it too is not a very easy plant. It is surprising that gentians almost indistinguishable from *G. pyrenaica* are found in the Pontic range of north-eastern Turkey, and I believe also in one area in the Balkans, but nowhere between. To return to the Val d'Eyne: as one ascends higher one reaches a zone of rather curious almost black rocky screes and here grows another treasure, *Senecio leucophyllus*, the white alpine groundsel distinctive in its very fine silvery-white divided foliage; the small yellow flowers are little addition. It is very striking in its own environment and probably better left there for it again is notoriously difficult in cultivation and should be tried only by the dedicated alpine gardener with an alpine house. It is close to the grey alpine groundsel *S. incanus* but more striking and silvery-white in its foliage. Unfortunately we did not find *Lilium pyrenaicum*, another Pyrenean endemic, which I would like to have seen in the wild.

The drive into Andorra was also interesting with plenty of gentians, the large flowered butterwort as in Kerry and some of the finest St Bruno's lily, *Paradisea liliastrum*, that I have ever seen, like miniature Madonna lilies and as glistening white. It is surprising that one sees this so rarely in gardens. We found it much pleasanter to stay outside the town in a little hotel on the mountainside rather than right down in the noisy and crowded valley.

To go west to Gavarnie involved coming back down to the main road and then going southwards later from Tarbes through Lourdes to Gavarnie. This is a tourist centre for pilgrims coming up from Lourdes to see the famous Cirque, gigantic and very impressive with tremendous dark cliffs and small waterfalls tumbling over. As long as one avoids the main path from the village to the Cirque where the tourist pilgrims amble slowly up and down on docile donkeys and mules, the other paths are fairly unfrequented and one is much more likely to see flowers there. It seems strange to find *Iris xiphioides*, the so-called English iris, growing abundantly in the meadows, but nevertheless it is native here. It flowers late, and the buds were only just breaking in mid-July to reveal the large exotic blue irises. The falls are variously flecked with white while the flowers have brilliant orange beards, a curious term applied to the stiff hairy crest at the base of the falls. The wild plants do not, however, have nearly as much flecking as the garden ones, in fact many are almost self-coloured, a very bright violet-blue. It is a bulbous iris of the same group as the Spanish irises and a good garden plant; in fact its vernacular name of English iris was given because it was first cultivated in England and thence introduced to European gardens. Who the hardy traveller was who first brought it from the Pyrenees to England is not known. Probably he was a seaman since, according to Dykes, it was first recorded towards the end of the sixteenth century as growing in gardens at Bristol, and Clusius was given stock from there and described it in his herbal.

Earlier there are dwarf daffodils such as the little *Narcissus juncifolius* in the meadows. A pleasant and very flowery walk takes one up to the great cleft in the rock ramparts separating France from Spain and known as the 'Brèche de Roland', since here, hard pressed and near to death after his unsuccessful foray against the Saracens of Spain, Roland is supposed to have rested and drawn his famous sword to cut a large cleft in the rocks, in which he broke the sword. *Fritillaria pyrenaica* also grows in this area although we saw it more plentiful on a pass farther to the west where it formed great patches. There it showed great variability in the colour of its bells, although we never found the pale yellow one, a great rarity. It is a dusky beauty, although the narrow bells inside have an ochreous green colouring. Outside they are dark slatish-purple varying in tone even in the same group. It is usually regarded as one of the easiest species to grow in English gardens.

In the woods behind Gavarnie on the rocks grow the ramondas, *R. myconi*, formerly better known as *R. pyrenaica*, and these are endemics of the Pyrenees. Although on both sides, we only found them on north-facing moss-covered rocks, their large rosettes of hairy rugulose leaves, like thick green corrugated cardboard, seem to hug the rocks, lying flat against them while the flowers in small clusters stand upright; these are deep violet-mauve and open flat to show the deep orange and yellow eye in the centre, and the stigma sticks straight out from this like a little spear.

37

We also found them in masses in the woods on the Spanish side of the western end of the range. There and also around Gavarnie was another magnificent endemic, *Saxifraga longifolia*, which makes great silvery encrusted rosettes sometimes six or even eight inches across. When the rosette has become large, probably after several years' growth, it produces vast plumes of white flowers, often two feet or more long. The plumes are massive with a hundred or more flowers, pure white with rounded petals and deep crimson stems, forming a large branched panicle. It is perhaps the finest of all the saxifrages and has the largest spikes of flower of any of its section. It is a plant of limestone rocks and usually grows in vertical crevices so that the moisture does not lie in the rosette. It also seems to have seeded fairly freely into the moss-covered stones of walls where it can grow undisturbed. After flowering it dies and does not usually leave behind any young rosettes, but it is readily grown from seed and this course should be followed for a garden supply.

One evening we met in the Hôtel des Voyageurs, that most delightful of mountain hotels, tucked away from the main tripper route at Gavarnie, Mr C. H. Hammer and his wife. He was a keen connoisseur of alpine plants and later became President of the Alpine Garden Society. Together the next morning we went down the hill a short distance towards Gêdre and thence eastwards along a wide valley where he had previously seen masses of a very fine dwarf recumbent form of the pink *Daphne cneorum*: and there it was, growing in large numbers among the short turf and in full flower, a most delightful little plant forming mats of bright pink stars sometimes nearly a yard across. It is quite prostrate and should probably be included as var. *pygmaea* although there seems to have been some confusion between this and var. *verloti*, but this latter is a taller plant with narrow leaves and looser clusters of flowers and it is probable that it does not really grow in the Pyrenees. The flowers of var. *pygmaea* seemed to be a slightly deeper pink than those of the type plant and it would certainly be well worth growing.

The Pyrenees are certainly an area that I would like to visit again since they contain a great diversity of fine plants and the mountains are very varied.

SPAIN

Late in May one year after the Chelsea Show was finished I set off for Spain with Eliot Hodgkin and Bertram Anderson and it proved a most rewarding trip. We aimed to visit three main mountain groups with differing floras, the Picos de Europa in Asturias in the north-west, the Gredos south of Madrid and the Sierra Nevada in the south near Granada.

The days of driving were long, since we only had a short time, but never-theless there was nearly always a flower detour to make a break. One of these took us off the road south of Bordeaux to see the giant Bayonne form of the hoop petticoat daffodil, *Narcissus bulbocodium*. It was certainly by far the largest form I have ever seen, standing over a foot high with a massive flower of lemon yellow on a stout stem. It was growing in dry patches of rather sandy soil among reeds, by now baked hard but earlier in the year it had probably been very damp. It is curious that this fine form has not been seen elsewhere and does not seem to have become established in English gardens as it is such an outstanding plant in its native habitat. The same seems to apply to some of the overpoweringly sweetly scented forms of the wild jonquils. Mr Hodgkin reports that the only place in which he can grow the Bayonne daffodil well is in the plunge-bed of sand in his alpine house. The Royal Horticultural Society gave his plant an Award of Merit under the name 'Bayonne'.

It seemed a long stretch along the north Atlantic coast where we drove steadily westwards until we turned inland for the Picos de Europa and stayed a few nights at Covadonga, a little town which has given its name to a lovely form of campanula. It proved a good centre. Whenever we stopped on the way we looked in the grass for crocus leaves with their distinct silvery midrib which always separates them from mere grass. We found quite a number, forms of the autumn-flowering *Crocus asturicus* and *Crocus nudiflorus*, both of which grow along this coast. It is a region of high rainfall and frequent mists, so that probably these two species do not need the complete baking in the summer of some of the Mediterranean species, and probably the same applies to the plants of the Picos. These are high mountains up to about 8,000 feet, and there was still plenty of snow on the higher ground in early June.

Other plants of the wayside were maidenhair ferns which grew in rock gulleys where it was damp and shady, and the giant butterwort, *Pinguicula grandiflora*, with its beautiful deep violet flowers each with a long spur. It also grew in damp places. Another excitement of the route was a short detour to see the famous cave paintings at Altamira, superbly preserved sepia and chestnut brown bison and slender hunters with spears.

Just above Covadonga we began to find daffodils in flower. This was *Narcissus pallidus praecox*, an early flowering trumpet with a rather ragged creamy-yellow perianth and a deeper yellow trumpet. It grew nearly always on sloping banks where there would doubtless have been good drainage, and was a pretty little plant. Higher up, again on sloping banks and sometimes almost in scree, we found our first *Narcissus asturiensis* and these were plentiful, scattered over quite a large area. This is the perfect little miniature trumpet only three or four inches tall which used appropriately to be known as *N. minimus*. The stems are weak and the flowers tend to droop on to the ground, but with their fringed edges and deep yellow flowers they are delightful. Obviously in cultivation this requires good drainage and does not need to be dried off too severely in summer. Here even in June it was cold with occasional showers of sleet at about 6,000 feet, and above us the snow was still quite plentiful. There was also a little *Fritillaria pyrenaica* still in flower.

Another interesting plant of this area, chiefly in the open woodland, was *Scilla lilio-hyacinthus*. This is unlike the normal Scilla since it has a bulb with numerous bare scales like a small lily bulb and broad shining leaves; the flower spike of pale lilac-blue flowers, which were not yet out, more closely resembles that of *Scilla italica* but is taller. At first, until we looked more closely, we thought that we had found small lily bulbs. It occurs also in the Pyrenees, where it has been described as taking the place of the English bluebell, although it is probably not so effective. In the more open places and in rock crevices grew pinkish and purple *Erinus alpinus* which always made a distinctive splash of colour, though we did not see any white ones. There were many of the more common alpine plants such as *Gentiana acaulis*, but few of these compared with the abundance of the Alps. There were also good forms of the white *Ranunculus amplexicaulis*, close to *R. pyrenaeus*, but with the stem leaves clasping the stem.

From Covadonga we went north-eastwards to the parador at Riano which we also found was an excellent centre although the hills around were not so high. It was a comfortable and pleasant place. The little hotels owned by the Spanish government and called 'paradors' were nearly always good places in which to stay. Near here were meadows full of the trumpet daffodil *Narcissus nobilis*, which is probably the largest of the wild trumpet species. It was indeed numerous and in full flower and made a lovely Wordsworthian spectacle blowing in the breeze. A curious feature of this species is the amount of deep yellow-orange at the base of the flower behind the perianth. It was also interesting to note that we saw a few that had the yellow streaks in the leaves symptomatic of one of the Narcissus viruses, and I had not realized before that this also occurred in the wild.

The Angel's Tears daffodil, *Narcissus triandus*, was also common, nearly always growing in very moist positions where the water drained down through the stony slopes. No doubt later it would be much drier but only for a fairly short period till the autumn rains and winter snows came. The little daffodils were variable, some forms having long cups almost as large as those associated with the variety *loiseleurii* from the Isles de Glenan off the west coast of France. These were tall, almost a foot high, with quite large clusters of drooping bells, creamy-white, some with a pale greenish-yellow flush. None was ice-white. These make lovely plants for a pan in the alpine house, but some have survived and spread by seeding for us on a bank in my woodland garden. The name Angel's Tears is said to have been derived from the name of a guide called Angelo who was employed by Peter Barr, the famous nurseryman who introduced this plant to England. After climbing for a long time he was very tired and burst into tears – and there was found this delightful little daffodil with its pendulous almost tearful flowers. There are other versions also of this story, and certainly the name is a most appropriate one and has remained. With it was *Tulipa australis*, sometimes even growing in running water – a habitat one does not associate with tulips. These, too, were beautiful. The drooping buds were deep yellow streaked with scarlet-crimson and when the flowers opened they became bright yellow stars.

Another interesting little daffodil grew on the steep screes in damp places and belonged to the jonquil section. The flowers were small with shallow cups and the narrow leaves twisting and prostrate on the ground. It was close to *N. scaberulus* but we had not heard of this species being recorded from this region. The hill slopes were covered with cistus and large shrubby heathers, *Erica arborea* and *E. australis*, although we saw no white form of the latter species. There were also rich pinkish-purple forms of St Dabeoc's heath, *Daboecia cantabrica*, but no better than those in general cultivation or those to be found wild in the south-west of Ireland.

From Riano we went south over the Guadarramas in a long drive to Madrid, and thence south-west again to Avila and the Gredos mountains in central Spain where we stayed at another excellent parador. Our most exciting sight on the way was a long hedge, perhaps a hundred yards or more in length, of the bright yellow *Rosa foetida*, growing along the edge of a park. It was covered in flower in a way that never seems to happen in England so that at first we were puzzled by it. It was a most magnificent sight since the flowers were quite large and the hedge was golden all over. I have only seen this rose flowering like that in one other place, and that was high up in the Elburz mountains in Persia where it grew round a little house owned by Count Hannibal of Tehran as a mountain retreat, and tucked away in a remote valley. Later the park superintendent kindly sent Mr Anderson a few suckers from Spain but they did not flower for us as they had done in Spain, lacking the hot summer ripening.

The Gredos were less precipitous mountains than the Picos, but were equally full of daffodils. Beside a little stream grew *N. pseudonarcissus*, our Lent lily, only in a rather larger form, while on the banks above and in crevices of the rocks was *N. rupicola*, fully justifying its name. This little jonquil, only a few inches tall, is a very pretty plant, larger in flower than *N. juncifolius* but with only a single flower to the stem, a good yellow orb nearly an inch across with wide petals and a very shallow cup, and very sweetly scented. But even more exciting when we climbed the steep rocky bank where grew the *N. rupicola*, we suddenly came out on to a vast expanse of meadow-like moorland, all yellow with the hoop petticoat daffodils *N. bulbocodium* in vast abundance and in varying forms. It was one of the finest daffodil spectacles that I have ever seen, far surpassing in area the alpine meadows at Wisley or in the Savill Garden at Windsor. There were various interesting species of *Linaria* also. It was a rich area generally, although unfortunately we did not have long there, but it was one to which I certainly would like to return.

Then we hurried southwards down to Granada and the Sierra Nevada. Granada was a delightful city with the old Moorish gardens of the Alhambra and Generalife palaces. These have remained much as they were in the fourteenth and early fifteenth centuries before the Moors were expelled from Spain and are still very beautiful. Like many eastern gardens they were designed for coolness and shade during the hot summer, with the constant sound of running water which the Moors were able to channel from the melting snows of the Sierra Nevada. Their courts were intimate and peaceful and private, more on the lines of a private garden today than the vast park of a palace. The hedges of clipped myrtle still border the canal and the jets of the fountains still meet like a ceremonial arch of swords at a military wedding over the long canal in the Patio of the Riadh, while the fountains in the Courtyard of the Lions still play. All this delicacy is contrasted with ancient dark and sombre cypresses. It is truly a magical garden, a series of green and flowery state-rooms in the open air but surrounded by Moorish arabesques of delicate architecture.

A long road from Granada leads directly up the north-west side of the great mass of the Sierra Nevada of which the highest points of Veleta and Mulhacen are over 11,000 feet. It was hot as we drove up to the university residence, a small and at the time rather primitive hotel, but there it was cold and patches of snow still remained melting around its doors even in midsummer, a delightful contrast to the hot plains below. This is probably the highest road in Europe since later in the year it is possible to drive up to the summit of the dome-like Pico Veleta, just over 11,000 feet, but in mid-June still deep in snow. The most exciting plant of our drive up was a little pink Viper's Bugloss, *Echium albicans*, with foliage like a mass of silvery and furry caterpillars and short spikes of very deep pink flowers, certainly the most delightful *Echium* that I have ever seen. We found it in piles of grit by the roadside left for mending the road and in hard gritty patches near. Unfortunately it does not seem to have adapted itself to cultivation in England. The northern slopes of the mountain are hot dry country covered with dwarf prickly bushes such as *Erinacea anthyllis*, formerly known as *E. pungens*, that superb little blue and mauve hedgehog of a broom. It was very variable in colour but everywhere equally spiny. *Daphne oleoides*, a small shrub with whitish flowers slightly flushed with pale purple, grew plentifully in the same area. In the gulleys were fritillaries, the dusky narrow bells chequered gold inside of *F. hispanica* and the narrow yellow trumpets of *Narcissus nevadensis*, both rare plants in cultivation. The daffodil is unusual in this section as there are several flowers to a stem.

Around the little residencia, even among the melting snows, were dwarf white ranunculus, *R. acetosellifolius;* the flowers were quite large on stems about six inches high, while the sorrel-like leaves spread flat over the ground. Among buttercups it is quite distinct in its foliage.

In gulleys a little lower the rare Spanish paeony, *Paeonia broteroi*, still flowered with tall stems and large globular flowers of pinkish-purple. There were also crocuses *C. nevadensis*, like mauve alpine crocuses flowering as the snow melted. There were yellow oxlips *Primula elatior*, very dwarf blue forget-me-nots and even a few plants of *Gentiana verna*. There are many more endemics on this range, while again there are different plants on the southern slopes which I have not visited. I would like also to be able to travel in earlier spring along some of the foothills and around Ronda to see some of the jonquils and other dwarf daffodils which grow there. It is a mountain range which merits a far longer visit, or visits at different seasons.

CORSICA

Our visit to Corsica was only a short one, but nevertheless it was very rewarding. My wife and I were lucky enough to be accompanied by three very knowledgeable gardeners, the late E. B. Anderson, a marvellous old man who always seemed to find as much as the rest of us although only walking a third of the distance; Eliot Hodgkin and Basil Leng of Antibes. We flew in early June direct to Ajaccio, the capital of this delightful island, and after some argument were able to hire a self-drive car and drive up the west coast road to Piana where we stayed at an hotel looking out over Les Roches Rouges, marvellous jagged pinnacles of deepest terracotta. While some of the coastal flowers were over, in the higher areas there were still plenty of cyclamen and crocuses and even paeonies.

The maquis was flowering with myrtle and rosemary and lavender and very spicy little shrublets such as thymes, while in the more acid areas there were tree heaths and arbutus. Scramblers such as dog roses and *Clematis cirrhosa* and prickly *Smilax* also abounded. The scent is so strong on a hot day that the islanders claim that one can smell the sweet scents of the island several miles out to sea. Coming by plane, of course we could not test this, nor was the weather by any means so uniformly hot and sunny; but still it was pleasant enough. The plant that most gardeners particularly associate with Corsica is the apple-green hellebore still known to most gardeners as *Helleborus corsicus*, but it has also been called *H. argutifolius* and *H. lividus* subsp. *corsicus*, but it is a much larger and more vigorous plant than *H. lividus*. It is still abundant growing along paths and at the edge of the scrub, and formed very large clumps often four feet across and nearly as tall. The best of the flowers were over and the beautiful jade-like apple-green of the fresh blooms was past but they still showed large clusters of fading yellowish-green beginning to droop over. One of the great merits of this plant is the very long season of flower. Its foliage also is almost as distinctive and decorative as the flower, a shining dark green with the leaf segments deeply toothed. In the garden it contrasts well with the large entire leaves of the bergenias. The white low-growing cistus *C. salviifolius* was the predominant species and lovely with its saucer-shaped flowers covering the bushes, while the yellow *Spartium junceum* clothed hillsides with bright yellow, as fine as anywhere I had seen it around the Mediterranean.

On the rocks right down by the sea grew another Corsican endemic, *Erodium corsicum*, a low creeping mat-like plant with geranium-like flowers varying from pale pinkish-purple, or even rose-pink, to quite a deep magenta-purple, each petal heavily veined with deeper colour. The later or the clearer pinks are perhaps to be preferred and the plant is an old favourite with alpine gardeners. The leaves are somewhat silvery with downy hairs and the whole effect is pretty. Like the hellebore, it is also reported as growing in Sardinia but nowhere else.

There were many terrestrial orchids; particularly fine were the Serapias, especially *S. cordigera* with its very wide deep maroon hairy lip. The greater butterfly orchid *Platanthera chlorantha* formed tall spikes nearly two feet high with large greenish-white flowers each with a long narrow spur. Both these are plants of the open places in the scrub. There was also *Orchis papilionacea* in varying forms.

Practically all the northern and central part of the island is mountainous and the little villages nestle between the rocky hills and the blue sea. I have seldom seen anywhere more lovely cascading displays of all kinds of pelargoniums, particularly the pink and red ivy-leaved ones, which fell from the balconies and walls of some of the older towns, and one courtyard has remained in my memory since for its beautiful masses of flowers.

We drove first along the coast, a very beautiful route, and in the sands around Isle Rousse searched in vain for the rare little pink snowflake *Leucojum roseum* which grows only in Corsica and Sardinia, but still it is autumn-flowering and so it was perhaps not surprising that we found no trace of it.

Like all Mediterranean islands, Corsica has its own endemic bulbs although usually sharing them with Sardinia. The little crocuses are very interesting forms of *C. corsicus* and *C. minimus*. We found these in several places, some in flower and some finished. They grew mostly in short grass among the rocks in the lower part of the mountains and the two species seemed to grade into each other and also into *C. imperati* from Italy which has larger flowers, pale buff on the outside with deep violet-mauve feathering and deep mauve inside shining as if lacquered. The botanical differences between *C. minimus* and *C. corsicus* lie in the tunics of the corms, that of *C. minimus* having parallel fibres while that of *C. corsicus* has reticulated net-like ones. The flowers of both are rather globular and very richly coloured, a pale lilac base colour being heavily feathered and marked with deep violet-purple which makes them some of the most beautiful of all crocuses even though the flowers are comparatively small. In the bulb frame I have found them fairly long-lived although in pots or in the open garden they seem to last a much shorter time. The increase by division is not very great and so they remain expensive in catalogues. In cultivation they are early flowering. With them we found leaves of small colchicums which were identified as *C. corsicum* when they produced flowers in the autumn, rather small but a good clear pinkish-lilac in colour. They are close to *C. alpinum*.

The big football-like bulbs and strap-shaped leathery green leaves of *Pancratium illyricum* were also plentiful and just an occasional white flower remained. Another plant which attracted my attention was the annual *Helianthemum* or *Cistus guttatus*, with numerous small pale yellow flowers, each petal with a dark maroon spot at the base. Often now it is placed in a separate genus as *Tuberaria guttata*, owing to its annual habit and lack of woody stems. Although it is widespread around the Mediterranean I have not seen it so well anywhere else. Another pretty little annual was *Sedum caeruleum* with the palest blue flowers only two or three inches tall and with slightly succulent leaves flushed with red.

Paeonia russi is an endemic plant to Corsica and Sardinia and a very lovely species with single large pinkish flowers which in cultivation appear

45

very early. We found a meadow in which it grew freely in the northern part of the central ranges, although only a little of it was still in flower. There are two varieties distinguished only by the hairiness or absence of it on the carpels. It belongs to the same group as the favourite garden plant *P. cambessedesii* from the Balearic Islands, but is larger. A small seedling that we collected has settled down well and flowers freely and even seeds around, but the flowers are not so deep in colour as those portrayed in Sir Frederick Stern's monograph of the genus.

After sampling some of the passes nearer to the sea we moved over into the central mountain area and settled at an old hotel at Vizzavona, just south of Corte and on the edge of the forest. Here it was cold at night and we were glad of the large log fires which burnt in the dining and sitting rooms even though it was June. It was typical mountain weather and even though only just over 3,000 feet up it was quite different from the coastal weather, cold and sometimes foggy at nights as the clouds came down. However, it was a rewarding area. Just south of Corte we branched off south-westwards along the Gorge of the Restonica, a wide valley ascending through forest towards the Monte Rotondo. There were many open rocky areas and here the deep pink *Cyclamen repandum* abounded in full flower. It is one of the most graceful of species and in English gardens the last species of the season in bloom. The petals stand upright, tall and slightly twisted, a lovely flower. At the base of the petals is a crimson ring around the mouth of the flower; up to an inch in length and carried well above the leaves, which are ivy-like. It is also strongly and spicily scented. I could never have imagined that we could have seen so many. Among them were occasional white ones and these must rank among the most beautiful of all the cyclamen.

Another day we found a path which went much of the way up to the refuge of Monte d'Oro one of the highest mountains of the area. Again we found cyclamen but not nearly so plentifully. But our main quest this day was the little white *Helichrysum frigidum*, another favourite of alpine gardeners. This grew on the rock slabs of the higher regions at about 6,000 feet and was quite common there. The little white heads are like fluffy balls of clear tissue, dry and crinkly to the touch and silvery-white, and borne on stems only an inch or two high above the thick mat of mossy rosettes. It is a perfect miniature but in this country needs usually to be grown as an alpine house plant. Another spreading plant of these rocks was *Prunus prostrata*, the smallest of all the cherries, but we did not see it in flower. Usually, however, this is disappointing for the flowers are small for the plant and inclined to be sparse in number. I doubt whether it would have been as good a form as Paul Furse and I found later high in the Elburz mountains of Persia. Another plant of interest from this mountain was the great white thrift *Armeria multiceps*, with stems about eight inches tall or even more, and round globular flower heads of pure white nearly two inches across. It was one of the finest thrifts that I have seen and it seems surprising that it is not in general cultivation in this country. We grew it for some years at the edge of a scree frame. Another unusual plant of the area was the white *Hyacinthus pouzolzii* sometimes known as *H. fastigiatus*, rather a dingy colour, though, with little black spots in the white. From Vizzavona we gently drove downhill to Ajaccio, the birthplace of Napoleon, and thence home again by plane after a very satisfying trip. Earlier we could have seen fields of white scented *Narcissus tazetta* down by the sea.

CHAPTER THREE

Greece and the Eastern Mediterranean

Stranger you have reached a famous land

. . . .

Here, day after day
Blooms the fair narcissus, fed on dew
The flower that crowns the goddesses
And here as well
Grows the golden crocus.

Sophocles, *Oedipus Colonnus*

The ancient Greeks were proud of their country and seem to have appreciated its beauty and its flowers. The study and enjoyment of the flowers combine well on a visit with a study of ancient Greek ruins, and several times we have been fortunate enough to travel with archaeological parties. Even so, the trip on which we probably found most flowers and relished the country most, was our first one, when my wife and I went by ourselves and were able to drive about and stop at will and had the thrill of discovering new plants which have since become familiar to us. Still the beauty of the country and the stunning abundance of the flowers in the spring never fail. One must go early though, unless one goes high into the mountains. The end of March and the first half of April are usually best. By the middle of May much of the flora in the lower areas is dry and brown, dormant and prickly until the autumn when the rains come again.

Greece has been compared to a natural rock garden and many of the areas are certainly the envy of gardeners. Early in April I have never seen such a display as the expanses of the little bright yellow *Trigonella balansae* like a bird's foot trefoil or a *Coronilla*, covering the grey rocks and running down to the dark blue sea of the island of Delos. The contrast of colour is magnificent. This is one of the finest areas for flowers, since the little island is uninhabited for much of the year and always only very sparsely. It is only a few miles long and across, and was sacred for hundreds of years to the cult of Apollo worship, with its slave-market and treasury of the Delian League, its lean and graceful panther-like lions and ruins of fine houses with mosaic pavements depicting dolphins and more realistic lions. There is a little mountain easily climbed, well under the hour, and with a view of magnificent expanses of the sea and islands of the Cyclades

48

spreading before one. Down by the shore there are pale mauve stocks with silvery grey foliage, mainly *Matthiola sinuata* whose leaves are distinguished by the wavy edge. These are perennials or biennials, but the more rosy-purple and rather dwarfer *Malcomia maritima*, the Virginian stock, is an annual. Both grow plentifully, especially near the sea, and with them grow the everlasting statices, *Limonium sinuatum*, also with grey wavy-edged leaves and spikes of deeper lavender-mauve flowers which will last dried for a very long time and retain their colour. These are, however, among the colours which seem to defeat the photographer and I have never been able to obtain a colour picture of the statice which satisfied me.

Further inland by the slave-market and lapping the bases of the stone lions are masses of a spreading pink annual campion, *Silene colorata*, a really bright pink, spreading like a carpet over the paving and contrasted with the infinitely variable deep purple or mauve Viper's Bugloss, *Echium plantagineum*, now sometimes known, perhaps more correctly but much less widely, as *E. lycopsis*. These are as fine as, and usually of a rather deeper mauve or purple than the seedlings from a nurseryman's catalogue. With the statice and campion grow masses of the mandrakes surrounded by ancient legends, *Mandragora officinarum*. The flowers are bell-shaped, a slaty-mauve or occasionally almost white, and nestle in a rosette of large and rather tough rugulose leaves spreading flat over the ground. This flowers early and by late April or May one will find the bright yellow seeds, encased in soft orange berry-like fruits, the size of small rounded tomatoes and softening to rather an evil-smelling mass. The legends relate mainly to the root which is thick and forked and vaguely man-like when pulled out. It was regarded as unlucky and even dangerous to pull this out oneself and so a dog was hitched to the root and if a scream resulted when the root was pulled only the dog was affected. Other legends relate to the potency of a drug derived from it for aphrodisiac and medicinal purposes, while in the Middle Ages it was said to have been used as a pain-killer during operations. I have not heard, though, of its being used as a drug of any kind in modern times, while as a garden plant it is rarely seen, but it is quite attractive and interesting nevertheless.

Then in other areas there are squirting cucumbers, another plant which the ancients used as a drug; asphodels, of which they ate the roots so that in German it is known as 'Potatoes of the Ancients'; delphiniums and lupins; golden yellow thistles, and other prickly scrub-forming plants such as the *Poterium spinosum* with its bright red bud scales, while everywhere else there are chrysanthemums, scarlet poppies and other annuals. Delos is famous for its flowers, and although many other areas have almost equal displays, early in the year it is an island of enchantment.

The eastern Mediterranean is the home of many of our best annuals but no display we could produce would exceed what we find here, where the chrysanthemums and bugloss and pink silenes have really run riot, for nature's riot is so much more attractive than that of the gardener.

The scarlet poppies and the yellow chrysanthemums as well as the Viper's Bugloss probably make the greatest contribution to the flowers early in the year. But one of the reasons why Greek flowers are so exciting to the botanist, be he professional or amateur, is the very high proportion of them that are endemics, that is plants which grow only in one particular area. Some have said that it is as high as one in ten. Many of these are restricted to particular mountains or islands. In many ways mountains and islands are rather similar habitats for producing endemic plants, those which have been isolated and then evolved on their own. For this reason some of the genera are very numerous, for instance there are nearly one hundred different campions and sixty mulleins. Few travellers can expect to know or find more than a small proportion of these. Some of the exciting genera for the gardener, such as paeony and cyclamen, fritillaria, crocus and colchicums, are richly represented and have varied tremendously from island to island, but more of these later in the next chapter. They have always proved the most interesting part of the flora for me.

Then the number of species of terrestrial orchids is very great, while within each species the plants are individually very variable. In some, such as *Ophrys*, which covers the bee and spider orchid as well as many larger ones that do not grow in Britain, two individuals even are not often quite identical. Their identification is very difficult and puzzling for this reason. Some have tended to split the species into a vast number of sub-species, others, perhaps wisely, to leave them grouped into large aggregate species. The bulbs have become particularly adapted to the very rigorous climate and short growing season. In many ways this is similar to the alpine environment in that they are forced by the climate into a short growing and flowering season and a longer resting one – only their seasons are reversed. Mostly they grow and flower during winter and early spring and rest during the hot dry summer from May till October; then they begin to grow again during the autumn rains, while some bulbous plants even flower then and produce their leaves later. In the lower areas there are usually no winter hazards of frost or excessive cold. The summer is the great hazard and the plants that grow most successfully are adapted to surviving this, either as annuals or as bulbs, tubers or corms, or by having reduced leaves and a prickly xerophytic structure, often with grey or hairy leaves, which help to reduce transpiration.

In the past the area has had a very varied geological history. As the ice ages of northern Europe retreated or increased and the waters of the sea, that was a larger Mediterranean, ebbed and flowed, some of the plants which did not like the hot dry summer of the lower coastal regions retreated up the mountain. So the flowers you will come across are as old as many of the rocks, perhaps the oldest things you will see – older than the remains of temples and treasuries and monuments.

Crete was probably joined at one time to North Africa, judging by the evidence of pleistocene fossils of mammals of the African type such as large and small elephants (*Elephas creticus*), dwarf hippo and deer, whose horns have been found in a shrine of the Snake Goddess. These animals must have wandered over a vast land mass, probably connected both with south Europe and Asia Minor.

Some areas were even separated from each other as islands in the surrounding sea. There the plants developed and mutated each on its own lines and new species were evolved. Mt. Athos is particularly rich in such plants. No goats used to be there. Evolution in plants has been particularly fertile in this region, perhaps corresponding in some way that we cannot begin to explain to the wonderful evolution of the human spirit that went on in Greece. Perhaps the bright sunlight and unusually clear air have had something to do with it. The Greek Flora, possibly like the people also, is a very polyglot one, a great mixture from many areas, some of which may have taken refuge there during the ice age. There are elements of northern European and Balkan plants mingled with more typical Mediterranean plants, while Turkish plants from the Middle East and the Steppes of Asia also come in, particularly in Rhodes and Cyprus which belong more to the Turkish system than to the western Mediterranean. I believe there is a deep water channel east of Crete between that island and Cyprus and Rhodes. After all, these two islands are very close to the Turkish coast.

There is unfortunately little true woodland or forest now left in Greece except on the mountains and even there it is only in belts. There was much more in classical and in Roman times all round the Mediterranean, but it was cut and not replanted by improvident people and the natural seedlings were mostly eaten by goats, that curse of all vegetation in the Mediterranean. They are the enemy of botanists too since they nearly always get there first and devour many of the more interesting plants. Even the ancients had this problem as is shown by this quotation from Eupolis of the fifth century B.C. which my colleague Mr Brickell, now Director of Wisley Gardens, discovered. It was a good rousing chorus of goats declaiming and there was not much that they missed. It is equally appropriate today.

On Arbutus, Oak and Fir we feed, all sorts and conditions of trees,
Nibbling off the soft young green of these, and these and these,
Olives tame and olives wild are theirs and thine and mine,
Cytisus, Mastic, Salvia sweet and many leafed Eglantine,
Ivy and Holm Oak, Poplar and Ash, Buckthorn, Willow and Heather,
Asphodel, Mullein, Cistus, Thyme and Savory all together.

The scarlet poppies of Greece seem to have a deeper and more intense red than those of England and this is accentuated by a black blotch at the base of the flowers. As one drives it is difficult to distinguish them from the anemones, and on our first visit when looking for anemones we frequently stopped to find that we were looking at poppies. When they have the light behind them, as in our cover picture taken near Ephesus by the Turkish coast, they glow with a brilliant and rich translucency that one finds in few plants. They are botanically, however, all regarded as forms of *Papaver rhoeas*, our common cornfield poppy; beautiful as they are here, they seem however, to be yet stronger in colour and larger in flower round the eastern Mediterranean, being only rivalled by the anemones and scarlet ranunculus. They are very widespread, especially near the sea-shore, and do not seem to be confined to any particular soils as long as the position is hot and sunny. With them frequently grow great sheets of yellow annual chrysanthemums, either *C. coronarium* or *C. segetum*, the parents of the strains of annual chrysanthemums of our florist. The flower of *C. coronarium* the Crown Daisy are slightly the larger and all leaves are deeply cut, while in *C. segetum* only the lower leaves are deeply cut and they are slightly bluish. This is the Corn Marigold which used to be found more widely in our cornfields but which has become much more sparse owing to the use of modern selective herbicides to the detriment of the colour of the landscape, but this is balanced by heavier and cleaner crops, so presumably the farmers do not regret it. There is also a paler cream and bi-coloured form of *C. coronarium* but it grows in equally large patches, occasionally mixed with the yellow, but more usually on its own. I particularly remember swaying vistas of these among the ruins at Knossos in Crete growing up against the ceremonial bull's horns, and also at Lindos in Rhodes among the pavements and pillars above the intense blue of the sea. Surely this is one of the most beautiful sites of the eastern Mediterranean. Just as common as the chrysanthemum is the little white daisy-like *Anthemis*, *A. chia* with the yellow disc centre, an annual only a few inches tall, very widespread on dry rocky places near the sea and in cornfields. The forms of *Campanula rupestris* are very variable among the ruins. They grow very well in the crevices of the old masonry blocks and may be seen in several of the ancient stadia and theatres, spreading starfish-like a yard or more across in the season and covered with mauve bells. The more usual form as seen at Delphi and Mycene has shorter bells and a less deep colour than the Rhodes form but is equally lovely. It grows also on Mt Parnassus and particularly well on the rocks around the Nauplion peninsula, which makes a good centre for visits to Mycene, Epidaurus and other sites.

The campanulas, poppies and chrysanthemums are particularly the plants one associates with so many of these sites, and wisely the Greek authorities leave them to make the ruins beautiful. They begin to flower in mid-April and go on through May. Unfortunately the campanulas are not hardy in England but can be grown well in alpine houses. They should only be collected as small seedlings or grown from seed since old plants make substantial roots right down among the blocks of stone. The campanulas of Greece include numerous endemic species both in the north in the Pindus range, on the islands and in Asia Minor. The little annual *C. drabifolia* is abundant on waste dry ground and also among the ruins. It has open deep mauve flowers smaller than those of *C. rupestris*, and is a hairy plant with rather lax spreading growth. Another campanula of slender growth but with larger, broadly funnel-shaped flowers of a strong violet-blue is *C. spathulata*. This is a plant more of open places in woodland or scrub. It is a perennial, but *C. ramosissima*, with more wide open violet bells and hairy stems and leaves, is an annual and grows particularly in the olive groves. The other little campanula which is widespread belongs to the genus *Legusia* which is separated by the segments of the flower being open and spreading rather than being bell-shaped or funnel-shaped. It is also an annual and is known as Venus's Looking Glass. The centre is white and the flowers pale violet-mauve, only three-quarters of an inch across but very numerous so that they make sheets of colour.

But the finest campanula of all is *C. ephesia* which grows in Asia Minor just around Ephesus and Miletus. It is particularly fine high up on the ruins of Ephesus and fortunately nearly always well out of reach, and I have found a telephoto lens helpful for its photography. The bells are very large like those of a Canterbury Bell and of a delicate mauve which looks well against the grey foliage. While the plant hangs down the bells turn upwards and outwards, so it is best photographed from above if possible. Again it may be grown in pans in the alpine house but it never makes the same effect as under the hot sun of Ephesus. I have never seen it anywhere else than in this small area and it seemed to be scarcer on my last visit to Ephesus than on the earlier one, but still one hopes that it will always survive. The authorities should nurture it perhaps and raise a few seedlings each year to plant out on suitable sites among the ruins, but only in its natural area. Behind the ruins are marshes with tall spikes of the yellow and white *Iris ochroleuca*, perhaps the finest iris of the eastern Mediterranean. They are also magnificent further south around Miletus and Perge.

The maquis of the Mediterranean, both east and west, is a low, and very aromatic scrub resistant to the drought and heat of the summer and nowhere else in the world does it grow like this. There are innumerable little thymes and strongly scented labiates such as mints, savoury, salvias as well as rosemaries. Two of the strongest in scent and commonest are *Satureia thymbra* and *Micromeria nervosa*, both with small purple flowers. The herbs grade into the small shrublets with a woody base. The thymes especially are used for flavouring meat and other dishes. Polunin and Huxley in their invaluable book *Flowers of the Mediterranean* suggest also that thymes are used with drying figs and prunes, but I have never been in the late summer.

These shrublets grade into the larger plants such as the evergreen Kermes oak, *Quercus coccifera*, which has leaves like a small holly and is named after the red *Coccus* scale insect which makes a little scarlet gall on the twigs and yields a fine red dye much used in classical times and probably even earlier. The acorns are rather scarce but sometimes one can find a few to demonstrate to the unbelieving that it is really an oak. All over the lower slopes of Mt Parnassus there are thickets of this oak, together with tussocks of the very prickly *Euphorbia acanthothamnos*. The euphorbias are very distinctive in the Mediterranean and give us some of the loveliest plants, especially when they are in flower and the low sun lights up the ochreous green of the flowers and bracts. The part that looks like a petal is really a bract round a very small flower. *Euphorbia acanthothamnos* makes a low-domed cushion of dense branches, deceptively smooth in its appearance, sometimes a yard across and about eighteen inches high. But it is not for sitting down and picnicking upon, for it is covered with green spines in a network like barbed wire and just as sharp. Even the goat does not plunge its nose into it although it will keep only a fraction of an inch above the spines and suck hard, thus getting many of the tender leaflets and flower heads.

I well remember driving up from the little port of Itea on the Gulf of Corinth to Delphi one late afternoon when the olives gleamed like silver against a dark thunder cloud and the euphorbias made unearthly greenish-yellow low domes in the landscape. There are very fine specimens also in the alpine moorland above the site at Delphi, a very rewarding area where the yellow asphodelines add to this predominant colour. Higher up we found dwarf irises, both the yellow and the purple forms of *Iris attica*, growing intermingled together, and in the woods *Anemone blanda* and, more rarely, *Fritillaria graeca;* while higher still near the Cave of the Winds were the black and green or often yellowish-green and brown flowers of the mourning iris, *Hermodactylus tuberosus*, which is sometimes offered in florists' shops. I enjoy nearly all the spurges but also among my favourites is the superb plant most of us call *Euphorbia wulfenii* although it has been variously named in authoritative books recently even by the same authority. *E. veneta* seems to be the most generally accepted, although volume two of *Flora Europaea* combines it under *E. characias*, to which it is certainly very close, as *E. characias* subsp. *wulfenii*, thus retaining the best known name. *E. characias* subsp. *characias* is distinguished by the black instead of yellow glands in the centre of the bracts, and is on the whole not quite such a vigorous grower and occurs more in the western Mediterranean, although it does reach Greece. These are lovely statuesque plants with blue-grey foliage and large domed heads, yellow flowers and rounded bracts. They are particularly fine in the Peloponnese. The other shrubby species is *E. dendroides* which grows larger than these, up to 6 feet tall and across and is a very handsome plant, though without the massive heads of *E. wulfenii*. It nearly always shows some red colour in the foliage.

More dwarf and spreading rather snake-like over the ground is *E. myrsinites* which has very glaucous and slightly fleshy stems and leaves, some of which usually turn red in the autumn. The flower heads are large and a strong ochreous yellow. The finest plants I have seen were in a pavement of a villa on the French Riviera, each a yard or more across with a radiating web of branches standing out against the grey stone, but it grows well all over the countries from Italy to Turkey. Close to it but a larger plant with more upright branches is *E. biglandulosa*, but it is too tender for most gardens in England. The arrangement of the pointed leaves on the stem has a distinct spiral twist which makes it easily recognizable. It is especially fine in the warmer more eastern countries such as the west coast of Turkey and the Lebanon, but does grow also in Greece.

Phlomis is one of the other abundant shrubby genera of the Mediterranean, in particular the Jerusalem Sage *P. fruticosa*. It is like a large yellow salvia with whitish olive-grey rather woolly foliage, the flowers appearing in whorls up a short spike. It grows into a good garden shrub in the warmer counties of this country, spreading into a four-foot bush and up to five feet across, and always looks well against grey stone – but then so does practically every plant. Among the more conspicuous flowers of the maquis and scrambling over dry banks is the caper, *Capparis spinosa*. The flowers are large, often three inches across with four rounded pinkish petals and numerous stamens with long filaments in the centre. From the buds capers are made for flavouring and pickling.

The three main woody plants of Greece have been long cultivated; the olive, so incredibly beautiful as it ages with its twisted trunks and silvery-grey feathery foliage; the cypress with its narrow spires so dark and rich a green, the vertical exclamation-marks in the landscape which contrast with the numerous horizontal growing bushes; and thirdly the vine, which requires no description. Sometimes one can manage to combine all three together in a photograph. These are integral to the Mediterranean landscape and do not seem to vary throughout a vast area. But the finest and generally the largest tree of the area is the Oriental Plane which lives to a great age and grows to a great size. It is reputed that the old trunk and some branches still survive of the plane under which Hippocrates, the great physician, taught in the island of Cos in 400 B.C., but I am doubtful over the authenticity of this. It is quite probable, though, that it is on the same site and may have been derived from a slip of the original tree repropagated at intervals. The leaves are large, palmate and with deeply divided lobes, a beautiful fresh green when they emerge in April while the young fruits hang like large round prickly baubles from the branches. The trunk is mottled like that of our London plane but massive old Oriental planes are nearly always finer than their hybrid children and it is surprising that they are not more widely planted as park trees. To sit under an old plane with a glass of retsina and congenial companions on a hot sunny day, as so many of the older Greeks do, is surely to savour the essence of Greece. Even more of them sit long over their glasses of ouzou. The other common deciduous tree is the poplar, so beautiful in the early spring, with its fresh yellowish-green foliage.

On sandy acid soils the arbutus and the tree heather grow, the latter more often a large bush with slightly dirty white flowers than a tall tree. There are two species of arbutus, *A. andrachne* and *A. unedo*, the strawberry tree of the south-west of Ireland. They can be distinguished by their leaf form, those of *A. unedo* having toothed margins while those of *A. andrachne* have a clear edge although a few of the young leaves may be slightly toothed. This differentiation can also be made difficult by caterpillar or other insect damage to the leaves of *A. andrachne* which makes them superficially appear as toothed. Both make lovely trees when mature, with peeling bark which leaves a mahogany under-bark. The flowers are white or slightly tinged pink, small clusters of drooping bells, and appear in late autumn and winter at the same time as the red strawberry-like fruits from the flowers of the previous season. Unfortunately they are very insipid and tasteless. Huxley and Polunin suggests that the name means

56

'eat one' and implies that one is enough. Certainly no one would be tempted to try more. In Greece it is plentiful in the sandy areas around Patras although more often making small shrubs than good trees, and I have seen finer trees in English gardens than ever in its native haunts. The same applies to the Judas tree, *Cercis siliquastrum*, so distinctive with its clusters of pinkish-purple pealike flowers and rounded leaves. There is, however, a fine and much photographed tree of this among the ruins at Olympia which flowers well in mid-April. The finest I have seen, though, are on the north side of the Bosphorus, they look magnificent against the dark hillside from the deck of a boat as one goes through the straits, especially in the low evening sun. Some of these must be very old since they are a good size and tree-like, while many are of a good deep pinkish-purple, deeper in colour than the usual form. I have not yet been able to travel on shore up the coast to see them closer but would certainly like to do so.

The white-flowered *Styrax officinalis*, the source of the resin storax, puzzles many visitors. It makes a large shrub or small tree and the flowers are pendulous. Otherwise it rather resembles a *Philadelphus* and the flowers are equally strongly scented. The resin makes incense and perfume and is said to be still used in some Roman Catholic churches. Each flower is nearly two inches across and they are carried in clusters or several on the undersides of the branches. It is quite common in southern Greece, Cyprus and along the west coast of Turkey.

Even more exotic in the landscape, particularly along the dried up beds of rivers, is the pink oleander, which grows into a large shrub and is abundant in places. It does not seem to vary in the wild and the double and white and crimson forms are only found in cultivation. As long as it can get its roots down to some residual moisture it seems not to be harmed by the summer droughts. Some of the finest bushes of it that I have ever seen, however, were growing on a seemingly very dry ridge near the road to Mycene and they were literally covered with flower. Even more strange and almost becoming naturalized in places, so that it has become one of the key plants one expects to see in the Mediterranean countries, is the South American Prickly Pear, *Opuntia ficus-indica*, of which hedges are made. Huxley and Polunin say that it was introduced by Christopher Columbus. Equally fierce and spiky is the statuesque *Agave americana*, a very distinctive plant in the landscape and also from South America. Then from Australia have come the different kinds of eucalyptus and of mimosa (*Acacia*); the one usually seen as a street tree covered with yellow bobbles of flowers is *Acacia cyanophylla*. Equally common as street trees are the Indian Bead Tree *Melia azerdarach*, with lilac flowers in spring and strings of bead-like berries in summer, and *Schinus molle*, the Californian Pepper Tree, while in really warm places one may find the lovely lavender-flowering Jacaranda from Brazil, surely one of the most beautiful trees in the world when it is in flower. One cannot travel far in the Mediterranean without realizing what a number of the important trees and plants, particularly around the towns, have been derived from overseas. Perhaps most widespread is the yellow *Oxalis* which seems to grow almost everywhere. It comes from South Africa and is known as the Bermuda Buttercup so it has invaded other countries as well. Perhaps it is at present the world's most efficient colonizer for regions where there is no frost, but it is nevertheless a beautiful plant with masses of strong yellow flowers – a real clear yellow. Occasionally one finds double forms and it seems as if the single form sports at times into the double and is still doing so.

The cistus is another plant typical of the Mediterranean region, and although there are only comparatively few basic species it is so numerous that often it forms the dominant plant in the scrub. They are lovers of the sun as are the helianthemums or sun roses and the flowers only open in the sun, but they do have a very long flowering season. The commonest is a rather dwarf shrub up to three feet or so tall with greyish hairy leaves and quite large magenta-pink flowers. The intensity of colour and the pinkness vary considerably. This is *C. villosus*, sometimes known as *C. creticus* or *C. incanus*, but these are really just different forms or other names for the same plant. The commoner dwarf white cistus in the eastern mediterranean is *C. salviifolius* which has leaves like a sage and flowers either singly or in clusters. There is also rather a dwarf bush not nearly so large either in growth or in flower as the Spanish Gum Cistus *C. ladanifer*. The flowers are a clear creamy-white with a yellow centre and cover the bush. There are two smaller flowered paler pink cistuses also, *C. parviflorus* and *C. skanbergii*, which I have seen particularly on Rhodes. The latter seems to be a naturally occurring hybrid from the parentage of *C. monspeliensis*, a white-flowered species, and the pale pink *C. parviflorus*. It originated in Greece and has spread.

A very odd and exciting plant which one sometimes finds among the cistus is the parasite *Cytinus hypocistis* which belongs to the Rafflesiaceae, the family which it shares with the *Rafflesia* of the East Indies which has the largest flowers in the world. In the *Cytinus* they are quite small and almost sessile on the ground, fleshy and light pillar-box scarlet from their overlapping scales with a yellow centre. There are no leaves and although a root parasite it seems to harm the host very little. I have never seen it in cultivation and it would be difficult to introduce unless one can bring its host also. It is, however, a startling find when one comes across it.

Another most decorative and common plant of the maquis is the French lavender, *Lavandula stoechas*, which, with the rosemary, gives some of the aromatic effect as one brushes through it. It has large purple bracts surrounding a head of dark purple flowers and grey-green hairy toothed leaves. It is probably slightly less aromatic than our ordinary lavender, but it is much more decorative, unfortunately it is not hardy in England except in very mild areas by the sea. It is a plant mainly of the acid sandy areas. The Spanish broom *Spartium junceum* is also very conspicuous, clothing banks of the roadside north of Athens, but the finest displays of it that I remember were in Corfu. The flowers are a very brilliant yellow and borne on the seemingly bare rushlike branches from late April right through to midsummer. I know of no yellow-flowered bush that makes a stronger effect in the landscape. It is a good garden plant also and in order to keep it bushy it should be cut over quite hard after flowering, but even then the bushes have a limited useful life, probably about ten years, since they become brittle when they get older. They are easily replaced, though, from seedlings. The ordinary brooms of Greece mainly belong to the genus *Calycotome* and are like very prickly gorse with golden-yellow flowers in abundance along the branches. The leaves are trifoliate but most of the terminal branches have become spines, while in *Genista*, some of which are also found, such leaves as there are have no divisions.

Small leguminous shrubs and sub-shrubby plants are also numerous, the commonest being the *Coronillas* and the *Psoraleas*. The *Coronillas* nearly all have glaucous foliage and bright yellow or orange-yellow pea-like flowers. As garden plants some of these, particularly *C. glauca*, flower intermittently throughout the year, particularly in a warm spot by a wall. *Psoralea bituminosa* is distinguished by its violet-blue heads of flowers and grows very commonly all round the eastern Mediterranean in dry open places. It is rather like a large pale violet clover and is a tall growing perennial, sometimes sub-shrubby at the base. It is one of the distinctive plants that one soon gets to know. The *Robinias* or false acacias are also abundant making medium-sized trees distinguished by their fresh green pinnate leaves and dropping white flowers. Natives of North America, they are now found in many countries as naturalized plants since they spread so freely both from suckers and by seed. They are widely planted by the roadside on banks and are useful in holding the soil from slipping or blowing away. They don't seem to suffer from damage by goats, the scourge of plant life in so many sub-tropical countries.

There are innumerable small vetches, peas, clovers, trefoils and astragalus and we cannot attempt to discuss them in any detail here. As well as looking at the flowers, a fascinating study lies in the shapes of their seed pods and by these many are distinguished. The Medicks are supreme in this, *Medicago marina* and *orbicularis* having seed heads like spiral snail-shells as does also *M. arborea*, a small shrublet with bright yellow flowers and silky grey branchlets and foliage. Others have seed heads like little spiny sea urchins, and in each species the form of seed head is different. To realize the full interest of their infinitely variable forms one needs to look at them through a hand lens. Then there are other little vetches and peas with inflated wings to the seed pods, such as the winged pea which has quite conspicuous dark red flowers and hairy grey green foliage. *Tetragonolobus purpureus*, as it is unfortunately named, does not have purple flowers but those of a good bright red and is easily distinguished by the broad wavy wings to the seed pods. It is an annual and very widespread. Then in *Physanthyllis*, such as *P. tetraphylla*, the calyx has swollen like a small balloon round the base of the flower. It is another annual and has silky grey-green foliage with rather large rounded leaflets and branches spreading over the ground. The flowers are pale yellow but tipped with orange and it is a distinctive plant. The annual star clover is another plant which is abundant and quite distinct, except that after flowering it hardly looks like a clover. The fruiting heads become dry and globular and the dried sepals become deep crimson, so that the whole resembles a little lantern covered with white hairs and deep red stars, each with a whitish centre. Another very distinct and beautiful trefoil is the prostrate *Trifolium uniflorum* in which the pale blush-pink flowers cover tussocks on the ground. It is unlike other clovers in that the pea-like flowers are single. It is particularly lovely on the slopes that lead down to the sea around Sunium, an area early in the year very rich in flowers, but one which dries up early. The ruins of the honey-coloured temple are so superb, standing high on a promontory over the deep blue sea, that the site is in any case always worth a visit. One friend claimed to have found over a hundred different plants there in little over an hour on one occasion in the early spring.

The acanthus is the plant that we associate with the Greek and hellen-istic decoration of Corinthian pillars, particularly the capitals at the tops where the stone is enlarged to support the roof, which was of wood and is now lost. The actual plant is usually *Acanthus spinosissimus* and the form of its great leaves, with their deep lobes so symmetrical and well balanced, is usually quite clear in the stone. Occasionally the leaves of the slightly less spiny and less deeply lobed *A. mollis* are used. It is curious that the very decorative spikes of flower with their hooded florets like a Greek warrior's helmet were never copied. It is quite common in dry places, flowering in late April and May, and I also enjoy it as a garden plant both for its form and its associations.

Bulbs, corms and tubers

In both spring and autumn these form for me the most interesting part of the flora and few areas are richer than the eastern Mediterranean. Let us start with the anemones. The two main species are *A. coronaria*, the poppy anemone of our florists and *A. pavonina*, the parent of our St Bavo strain and a more graceful slenderer plant. They are quite easily distinguishable since the poppy anemones have finely divided leaves at the base like a fern frond and also finely divided bracts on the flower stems, while those of *A. pavonina* are deeply lobed but not fern-like and the bracts or stem leaves are only divided into three, although each lobe may be split again near the top. Both species start to flower early and in Greece are usually at their best in the latter part of March and early April. There are wonderful patches of the scarlet form of the poppy anemone beside the road from Athens to Sunium, a brilliant dazzling scarlet but tempting to the local boys for picking and selling to tourists. One usually does not find the forms mixed and in other places one finds mauvy-blue forms and more occasionally white ones. Some of the scarlet ones, however, are white towards the centre while others are solid red all through with a glorious silky sheen. Usually they don't have more than six to eight petals and are two to three inches across, while those of *A. pavonina* are borne on taller and more slender stems and seem to grow in slightly moister positions. Their flowers have eight to twelve petals and, apart from the solid scarlet one which is perhaps the most beautiful of all anemones, there is a form with a white centre which has been distinguished as var. *ocellata* and one with purplish-mauve flowers which has been called *purpureo-violacea*. I remember first finding this in ditches beside the road to Delphi, while the solid scarlet form was growing freely under the olives by the road from Delphi down to Itea, especially on the edges of the little ditches. Some of the scarlet forms are very close to the commonly cultivated *A. fulgens*, but this generally has a few more petals and is thought to have arisen as a natural hybrid between *A. pavonina* and *A. hortensis*, a plant with white or very pale lilac flowers, more starry and generally less robust. It is more a plant of the western Mediterranean but does extend to western Greece. There has, however, been much confusion and many changes in the nomenclature of these common anemones. These are plants of the lower areas, and the mountain anemone of Greece is the little blue-mauve *A. blanda*, a lovely starry flower only a few inches tall. From Greek forms of this has been selected the beautiful deep mauve form called *ingramii* or *atrocoerulea*, which has proved such a superlatively good garden plant, spreading into a carpet of colour on the warm chalky soils in some parts of the south of England. We have seen it in the patches of spruce woodland on Mt

Parnassus and also in the open rocky ground around the grey temple of Bassae in the Peloponnese. Here it was snowing slightly in early April.

The derivation of the name anemone is in doubt. The popular and seemingly very appropriate one is 'windflowers', from the Greek word for wind, but other authorities have derived it from the Syrian 'Na-ma'an', the annual cry of lament for the dead Adonis whose blood is described by legend as flaming yearly back to life in the flowers of the scarlet anemone. Nevertheless, it is one of the flowers we associate particularly with Greece.

The cyclamen of Greece are also very widespread and one sees fine masses even on Mt Lycabettos in Athens itself, while there are expanses beside the road going up to Mt Parnes – a rewarding day or even after-noon trip for the flower lover. These are the autumn-flowering *Cyclamen graecum* and *C. neapolitanum*. *C. graecum* is the finer species with more rounded leaves with a slightly horny margin and a more rounded tuber, but both are infinitely variable in the markings of the leaves. It is, however, too tender for cultivation outside in England except in very warm gar-dens in the south, while *C. neapolitanum* grows freely in most gardens. The flowers vary from pure white to pale pink and those of *C. graecum* always have a deeper purplish crimson patch at the base of the petal and are slightly larger. They are most graceful, like little skirted ballet dancers, and the flowers come in early autumn before the leaves appear or with the first young ones. There is no plant that make such good ground-cover as *C. neapolitanum*. Curiously in this species the roots all emerge from the upper surface of the tuber, while those of *C. graecum* come from the centre of the base. If one wishes to collect a few tubers it is usually much better to take only small ones since these establish much better than the very big old almost woody ones which tend to rest for a year or two after being out of the ground for long. They may be surprisingly deep down and it is important not to break the small branch or even thread-like connections between the leaves and the tuber.

The only spring-flowering cyclamen that I have seen in Greece is *C. repandum*, which was flowering in mid-April in the Peloponnese but not nearly so abundantly as we saw it in Corsica or northern Italy. The flowers are pale carmine-red and the petals long and graceful and usually slightly twisted but with no blotch at the base. The tubers are much smaller and rounded. It is more a woodland plant and grows better also in English gardens in slightly shaded places. All these are wonderful plants for growing in pans in the alpine house and very little trouble. *C. persicum* has been recorded from the south of Greece but I have not seen it there. The Mediterranean islands nearly all have their own particular cyclamen.

The crocuses of Greece are numerous, the majority being mountain plants. The finest display I have seen was on the slopes of Mt Chelmos, in the north Peloponnese in mid-April. This was the three-coloured form of *C. sieberi* called var. *tricolor* and grew among small rocks and scree, flowering very soon after the snow melted. The flowers are deep mauve and quite large and have a very distinct yellow throat above which is a broad white zone, thus making the three colours. We even found one that was completely white except for the yellow throat like the one in cultivation known as 'Bowles' White'. With the crocus grew the very strong blue *Scilla bifolia* and the bunch-flowered spring-flowering *Colchicum catacuzenium*, with pale rosy-purple rather globular flowers each nearly as large as the crocus. It grew under very damp conditions in scree running with water from the melting snows and is more difficult to establish in English gardens. This was one of the most rewarding excursions that I have made in Greece, though it seemed a long trek from the village of Kalavrita where we stayed, a sad little village only just rebuilt after savage treatment from retiring German troops.

The more normal form of *Crocus sieberi* is var. *atticus*, which is deep mauve except for the yellow throat and is fairly common on Mt Parnes and in other hilly areas of northern Greece. The other form, var. *versicolor*, perhaps the most beautiful of them all, grows only in Crete in the White Mountains. The golden-yellow crocuses of the mountains are mostly forms of *Crocus chrysanthus* which has yielded us such a marvellous range of early-flowering garden varieties, but we have never found them very abundant in Greece. *Crocus aureus*, probably the finest orange crocus of all, the flowers being tinged with reddish-orange over the deep yellow, is an early spring flowerer in northern Greece, starting in January. It is the parent of the widely grown 'Dutch Yellow' but a finer plant, more brilliant in colour, with slightly larger flowers and with the advantage of spreading by seed which its sterile progeny does not. I have never seen the autumn and winter flowering crocuses in flower in Greece although they are well known in our garden. These include *C. laevigatus* with pale lilac coloured flowers heavily flecked with deeper mauvish-purple and flowering from November till February, *C. sativus* and its variety *cartwrightianus* with lilac-purple flowers and large orange stigmas in autumn. These flower best for us in cultivation in the bulb frame where the corms are warm and dry all the late summer. *C. cancellatus* has pale lilac smaller flowers than in *C. sativus* but heavily feathered with deeper purple, and flowers from late September onwards.

I have never found the beautiful *Crocus boryi* of the Peloponnese since I have not been there in its autumnal flowering season, but I have flowered it in my bulb frame in November and it is without doubt one of the loveliest species, a pure white with deep mauve flecking round the base of the flower. It cheers me also for it brings to mind the account of Greek flowers in the first of the Royal Horticultural Society's Journals which I helped to put together. It was by an old friend, Mark Ogilvie-Grant, who had lived long in Greece and knew its flowers and exactly where to find each of them better than anyone else I have met. During the war he returned to Greece for special work and was captured. He records that, while awaiting trial as a suspected spy at Tripolis in the Peloponnese, during exercise on the neighbouring hillsides, he was able to borrow the bayonet of his guard, described as 'an amenable companion', to dig up corms of this beautiful white crocus which grew with the yellow *Sternbergia lutea*. It is true that a real love of flowers and plants never deserts the true disciple even under the most difficult and anxious conditions. However, Mark survived to tell this tale and also to guide us on Parnes to find the rare *Celsia boisseri*, a verbascum-like plant a foot or more tall, and with little black bees in the centre of each flower. It would indeed be a nightmare to dream that one's interest in plants might ever leave one.

The majority of the colchicums also are autumn-flowering. The commonest is *C. sibthorpii* which has very large pale globular rosy-purple chequered flowers, but in spring they are only recognizable by their leaves, not unlike those of the common garden *C. speciosum*. The sternbergias, with flowers like bright yellow crocuses, are also autumnal, and in gardens the narrow-leaved forms from Greece seem to be the most free-flowering. Both these and the *Crocus sativus* are said to grow even on Lycabettos in the middle of Athens with the cyclamen. Our only visit to Lycabettos was at Easter when towards midnight we climbed in a great concourse up to the little church at the top, each clutching an unlighted candle. On the stroke of midnight the priest dispensed the sacred fire to those nearest, who passed it on to the others until there was a procession in the dark of little candles winding down the rocky path. If by mischance the wind blew out one's candle one must seek to relight it from another candle, and all seemed generous in distributing their little fire in this way. An unknowing visitor who took out a cigarette lighter to relight hers was regarded with sour looks and disapproving murmurs which I expect she will not quickly forget.

The fritillaries and the tulips are the two other groups of bulbs which I remember particularly in Greece and it was always exciting when one found one for they are occasional and often solitary and so should be left where they are. On Mt Hymettus we found the occasional specimen of *Fritillatia graeca* and saw these again, growing slightly larger, in the patches of spruce woodland on Mt Parnassus, but they were never common. The

FLOWERS OF EASTERN TURKEY AND IRAN

(*Above left*) *Iris acutiloba* has strongly marked flowers of chocolate-brown and grows in Western Iran on the mountains. (Photo: Paul Furse).

(*Above right*) *Iris urmiensis* has bright yellow flowers and grows around Lake Rhezaieh in Western Iran.

(*Below*) *Iris iberica* by evening light outside Erzerum in Eastern Turkey.

PITCHER PLANTS FROM BORNEO

(*Left*) *Nepenthes ampullaria* in which the red and green pitchers hang round a liana-like stem.

(*Right*) *Nepenthes rheinwardtiana* which has long crimson pitchers, jade-green inside with two crimson spots which attract the insects.

bells are a dusky rich brown chequered with jade green and usually with a rather broad but ill-defined green band down the centre of the petals. Inside, the jade is mingled with the chocolate brown and a rather dark dusky gold. The leaves are glaucous. In northern Greece the main fritillary is *F. pontica*, a taller plant with larger bells of jade or yellowish-green overlying a purplish-brown. The outer petals have a broad greenish stripe down their centre.

Of the tulips, the large scarlet *Tulipa boetica* was being sold in bunches in the streets of Athens during our first visit although we never saw it there again in such quantity. We did not see it growing near Athens although presumably it does so. We did, however, find it growing in a cornfield below Delphi, the flowers, although nearly a foot tall, being hidden from a distance by the young corn. It was only by chance that we found this particular field where they were plentiful. It was not possible, however, to photograph them as they grew, without cutting a large patch of corn. This we did not wish to do and so had to take them in a vase on the balcony of our hotel. They flopped over the side of the vase though in an unusual way, like over-forced cut tulips sometimes do when put in a hot room. The flowers are nearly three inches long and a bright pillar-box scarlet with a well defined dark olive-green blotch margined with yellow at the base of each petal. They look like a small Darwin tulip. The bulbs were about ten inches deep in the heavy clay, probably below the level of the very shallow plough used and hence they are able to survive. The other tulips which we saw on Mt Parnes, although only rarely, were the widespread *T. australis* and the larger dusky bronze and buff-orange *T. orphanidea* flushed with green and purple. More often, though, one only found single leaves of young bulbs rather than flowering plants. The Greek forms of *T. orphanidea* are not, however the equal of the magnificent and largest flowered Turkish forms from the area of Smyrna which have been distinguished as *T. whittallii*.

A discussion of the orchids I will leave to the next chapter on the islands where I have seen them more plentiful and finer than in Greece itself, although on our first visit we did find superb *Ophrys mammosa* quite plentiful in the ruins of Delphi, and the gigantic but rather coarse *Himantoglossum (Barlia) longibracteatum* in the woodland above the stadium; but in subsequent visits these, alas, have been very much rarer and may have been over-collected. They should now be left where one finds them. Few of them make good garden plants in England and are even difficult to keep long as alpine house plants.

Of the true bulbous irises there are probably none in Greece although the xiphiums recur in the western Mediterranean and the reticulatas and some of the Junos to the east in Turkey and Iran. Greece seems to have been left out between the two. The very variable forms of the tuberous *Iris attica*, however, make up for the other omissions, while there are two tuberous and corm-forming iris relatives which have both been placed in the genus *Iris* in the past and are often still described as irises. The first is *Gynandriris sisyrinchium*, formerly *Iris sisyrinchium*, which gives us such beautiful blue starry flowers from the sand and short grass all round the coast. These are plants of the sun, only opening in sunlight and even then for a very short day, when they look like flocks of little blue butterflies. When closed they look just like grass or a rush and will rarely be recognized. The rootstock is a little round corm covered with a dense highly reticulated network. Again, they do not often flower well in England although I have had the occasional flower in our bulb frame. The other is the mourning iris, *Hermodactylus tuberosus*, whose flowers are sometimes sold in florists' shops. Again they are very variable. Our first experience with them was on Mt Parnassus above Delphi where we found them growing in the moorland around the Cave of Zeus. I still remember what a perishing icy wind blew there in early April, but it was well worth it.

The dark blue *Muscari commutatum* are common around Delphi and probably also in other areas, just like small grape hyacinths, although much darker than the 'Heavenly Blue' grown in gardens and rather a menace among choicer plants. The spike always has a few sterile flowers at the top which never open properly and these are of a paler blue giving a slight Oxford and Cambridge effect, although not nearly so much as in some other species from farther east. The more abundant tassel hyacinth, *Muscari comosum*, is a larger stouter plant and common all round the Mediterranean. In this the top sterile florets have elongated, like bright violet-blue tassels. Otherwise in flower it is a dull rather ochreous brown.

CHAPTER FOUR

Islands of the Eastern Mediterranean: Crete, Rhodes, Cyprus

Crete, Rhodes and Cyprus are the three islands about which I intend to write in this chapter, and our travels there, although all too brief, include some of our richest flower-hunting experiences. Most of the commoner flowers that we have described under Greece are there but in addition there are so many endemics, some of great beauty and horticultural value.

CRETE

Crete has some of the most magnificent scenery in the Mediterranean, it has a warm and pleasant climate, not too dry in spring, and in addition has its own particular cyclamen, paeony, tulips, crocus, anchusa, arum, ranunculus, palm, the finest clover I have ever seen and several other interesting genera as well as masses of orchids. All these are quite distinct from the mainland plants and also from those of Rhodes and Cyprus from which it is separated by a deep water channel. There are two main centres which serve well as bases; both are ports and situated on the north coast, Heraklion in the east and Canea in the west. Heraklion is the centre for the Minoan archaeological sites of Knossos and Phaestos and also for excursions to Mt Ida, which has a large snowcap in April; while Canea is a good base for visits to the White Mountains, the plain of Omalo and the spectacular Gorge of Samaria. The area to the east around Agios Nicolaios is also an excellent base for flowers and from there one can visit the upland plain of Nidha and its famous Cave of Zeus. Crete was indeed a much God-haunted island for the ancients with its spine of mountains along the centre and numerous caves, reputed birth places of the gods.

The cyclamen, *C. creticum*, is a delight and on a recent visit in early April I was surprised how abundant it was; later towards the end of April and May the flowers are very sparse, only among rocks in the higher places. It is mainly a plant of the open woodland and shady banks in the lower areas. The flowers are white with long and slightly twisted petals of the *repandum* type and it is distinguished by a strong, rather acrid spicy scent. The leaves are ivy shaped but variable, pleasantly

67

marbled and with a rather deeply toothed edge. In cultivation it is a little tender but I have grown it outside in Sussex in a slightly shaded wall. It seems to do better out of the full sun and I have known one authority on the genus who grew it, together with *C. libanoticum*, most successfully under the bench of his cool greenhouse. In a genus of beautiful plants I think it is one of the most lovely. I particularly remember one old stone wall in open woodland between Heraklion and Canea where the largest specimens, some with flowers nearly six inches tall, grew out of every crevice. The tubers are rather small, smooth and round like a large well washed pebble, and usually five or six inches deep in the soil. We saw it again, rather more dwarf but flowering equally well on ledges near the upper part of the Gorge of Samaria under the shade of small pines and cypresses, a wonderful setting.

The paeony *P. clusii*, formerly known more frequently as *P. cretica*, is one of the most superb species of this magnificent genus. The flowers are large, rather globular and either pure white or ivory in colour. I have even seen some with a faint yellow tinge, but the usual ones are white, four inches across when open, broad petalled and with a large boss of golden stamens and purplish-crimson carpels in the centre. The base of the flower is slightly flushed with crimson while the stems, a foot to a foot and a half tall, are deep crimson. The leaves are deeply dissected, almost fern-like. It grows rather locally in the west of Crete in the White Mountains around the plain of Omalo, and in the Gorge of Samaria, and occasionally back from the road crossing the island from Canea to Sphakia, a little port on the rocky south coast and one of the most charming villages there. The way down is a fearful zig-zag descent but it is relieved by the masses of deep ochreous yellow spikes of *Asphodeline* in early April. I have not seen this so plentiful elsewhere. The slightly greenish tinge to the deep yellow of the flowers seems to contrast well with the grey rocks.

The plain of Omalo at nearly 4,000 feet above sea level, south from Canea, is one of several examples of these high saucer-shaped and very fertile plateaux which seem to be peculiar to Crete. The plain of Lasithi in eastern Crete is another, and this one is covered with windmills to draw up water, their white sails perpetually revolving in the wind, a fascinating sight. This is one of the largest upland plains, several miles across. One guide book claims that there are ten thousand windmills on the plain. On the north-east slopes of Mt Ida, the highest mountain in the island, is the Nidha plain, the third of these very peculiar areas. They are damp in the spring, with water draining down through potholes in the limestone from the surrounding mountains.

The first time we went up to the plain of Omalo we took an ancient taxi from Canea and found the upper part of the road very rough with vast potholes. The loquacious taximan was still full of the misdeeds of the German occupation force, a subject which after the war not unnaturally obsessed many Cretans, particularly in the west. One drives through the most magnificent groves of oranges and the sweet scent of the flowers was borne on the wind while the branches were still covered with fruit. Here are reckoned to be some of the finest orange groves in the island. At the last village we picked up an upland Cretan farmer who was going up to the plain to start his crop, and who chatted in broken English with my wife while I looked for flowers. It was early April and an icy wind blew over the plain, which is several miles across and was surrounded by still snowy mountains. Although heavily cultivated, it is one of the finest areas for flowers that I know. Some of the fields were full of *Tulipa bakeri* a most delightful species with good-sized flowers of bright pinkish-purple or even deep pink, each with a deep yellow egg-like centre surrounded by a white zone. It is distinguished by larger stems having two or three flowers, a rather unusual character in tulips. It is a most unusual combination of colours, rather striking but not one to mix with other tulips. The bulbs survive since they are below the level of the rather shallow ploughing. If ever the Cretan farmers decided to import more modern deeper ploughs the bulbs might be doomed; by so slender a margin do some species survive. However, a few of the tulips would probably remain in the banks between the fields but the population would be much diminished. The other Cretan tulip, *T. saxatilis*, is earlier-flowering and rather similar in colouring, a little larger in flower but paler, pink tinged with a lilac-magenta and again with a deep yellow centre. Both of these are distinguished by their shining fresh green leaves and tend to spread through stolons. In cultivation they do not usually flower freely unless they can be root-restricted and given a strong baking in the summer, when presumably the soil of their native habitat gets dry and hard as a rock. There is another Cretan tulip of the same group, *T. cretica*, a much smaller plant with white flowers flushed purplish-red outside, practically no stem and narrower leaves. It is rare, however, and I have only seen it once quite high up in the White Mountains but it has been recorded also from lower down, flowering early in the season. It has only rarely been seen in cultivation.

Apart from tulips there are wild arums on the plain. *Arum creticum*, is a variable plant, in its best forms a rich bright yellow, a little larger than our common Lords and Ladies, in its poorer forms a rather dull ivory white with a cream tinge. It grew there in patches among the young potatoes which were beginning to come up and among the prickly berberis. It is a good alpine house plant and hardy in warmer gardens.

When we first went up to the plain one had to walk from the end of the road as one entered the plain, across to the top of the Gorge of Samaria, but now there is a good road, firm at all seasons and even a tourist pavilion at the end of the gorge. At the head of it we sat down for lunch on our first visit on some short grass and among it were narrow crocus leaves, easily recognized from close to by their narrow silver strips down the centre of each leaf, and quite clear once one has seen the first, when one quickly finds more. This was a plant I was particularly anxious to find, the Cretan variety of *Crocus sieberi*, and in fact the first to be named, thus entitling it in strict botanical priority to the name *Crocus sieberi* var. *sieberi*. It is far more often, however, known as var. *versicolor* or var. *heterochromos* and both these names describe it well, in fact saying the same thing, one in Latin, the other in Greek. It flowers particularly early and these were long over, although higher up it has been found in flower as late as April as the snow melts. The basic colour of the flowers is white or a pale lilac-mauve but they are feathered and blotched towards the tips of the petals with deep purplish-maroon, in absolutely infinite variety so that hardly two individuals are quite identical. At the base of each is the solid golden-yellow throat so significant of this species and the stigmata are long and deep orange-scarlet. Some have called it the most beautiful of all crocuses and it would be hard to dispute its claim. The leaves are much narrower than in the other varieties and the corms are smaller and unfortunately it is a much less vigorous plant in cultivation. However, it flowers for me each year in my bulb frame in February, undisturbed and baked all the summer.

The Gorge of Samaria is one of the greatest natural features of the island. About nine miles long, it descends in a deep cleft between the grey craggy and very sheer precipitous mountains; sometimes one can walk on a sloping and slippery stone path, at other places a staircase of stone has been made. The Cretans describe it as 'zig-zagging down the Xyloskalon', their name for the rock staircase. In some parts water runs down it in a little stream which can become a torrent after heavy rain or much melting of the snow above. It is one of the last haunts of the famous Agrimi, the Cretan mountain goat with superb horns, but I have never been fortunate enough to see one and they are now very rare. We did, however, see hawks and eagles and even one of the vast lammergeirs, soaring over the peaks. It must be one of the wildest parts of Europe and so steep is the rock that probably it can never be spoilt or changed much, nor is it easy to imagine anyone wanting to do so. In clefts of the rock cling very ancient, gnarled and twisted and infinitely wind-blown cypresses and a few pines. The cypress is another Cretan endemic being the horizontal rather than the vertical form of the Italian cypress, and it is quite distinct and very decorative. In smaller clefts of the limestone rock and on shallow ledges *Cyclamen creticum* grows, also a yellow shrub-like flax *Linum arboreum* and golden drop *Onosma erectum*. This is a delightful plant with very bristly grey stems and leaves and a spike of pendulous deep yellow drops sometimes tipped with a little orange. It is very close to the slightly larger *O. frutescens*, but this makes more of a woody plant at the base and has more reddish-orange in the flowers. The best way to grow these in England is in crevices of a dry wall facing south, as warm and dry in the winter as can be managed; then they are lovely plants and build up into large clumps.

A fine excursion is to walk right down the gorge to a little town, Agia Roumeli, passing the little village of Samaria, from which one follows the river bed downwards. It is sometimes possible to arrange for a boat to come and pick one up at the harbour below Agia Roumeli to take one eastwards to Sphakia or westwards to Palaiokhora whence one can get a bus back to Canea, but I have not done this. Alternatively one can walk to Sphakia but it will take several hours, and it is an even more arduous walk back again up the gorge to the tourist pavilion.

Instead of going down the gorge from the tourist pavilion one can mount towards the peaks, and this is equally rewarding and the views down into the gorge are very striking. Here grows the wild forerunner of our common aubrieta *A. deltoidea*, with masses of small mauve flowers, while higher up are groups of little gentian-blue chionodoxas, *C. cretica*. These are smaller in flower than the commonly grown *C. luciliae* but a little taller and looser in growth and have a conspicuous white throat. Higher still one may find more crocuses, in flower later than down below as the snow melts, also small colchicums, but the most interesting and spectacular find will undoubtedly be the cushion-forming *Anchusa caespitosa* if one is lucky. It is a plant of rock crevices and has narrow strap-shaped leaves and the very brightest of little blue starry flowers, almost sessile on the cushion. It makes a treasured alpine house plant.

Another very satisfying drive is from Heraklion over the spine of the island to Phaestos, where the ruins of the Minoan palace are most beautifully placed looking out over a vast panorama to the south and east. It is a far finer position than the site of Knossos which has no view and the sea has receded leaving it a few miles inland on the coastal plain. On this route I have found the cyclamen in the banks of muddy thick clay, like plasticine after rain in the spring, while there are also some of the finest masses of the white form of *Ranunculus asiaticus* which is numerous here and an endemic of Crete. Further east around Agios Nicolaios there are some groups of the yellow form, but this really belongs to Cyprus. The white form is a most beautiful plant, graceful as it blows in the wind. Each flower is about two to three inches across, wide petalled, a good white and with a black centre. Some are flushed with pink and are worth special selection. The ranunculus is easily distinguished from the anemone by the flowering stems which are clear of leaves or bracts while the central boss of stamens is more dome-shaped. The leaves are also less finely divided than those of the poppy anemone. Further east still, in Rhodes and along the Turkish-Lebanon coast one finds the scarlet form, which is perhaps the finest of all the forms. These are the parents of the Turban ranunculus of florists, but in the wild the single forms give me more delight than the more gaudy double ones.

Along the banks beside this road may also be seen the giant clover of Crete, *Ebenus creticus* and it is a most striking plant, sub-shrubby with finely divided very silvery foliage and large conical flower heads of a good clear pink. Sometimes the clumps are a yard across and clearly very old, making a dome of stems two foot high. The flower heads are about two inches long. It seems to be local in Crete but in places there are large stands of it, often growing in clefts of almost vertical walls. It is particularly fine almost opposite the entrance to the Minoan palace of Agia Triadha near Phaestos and also in a wall beside the road a few miles beyond Knossos as one goes eastwards. As a garden plant it is largely untried, but a cutting from a small seedling I collected once in Crete has survived the last winter in a wall facing south in Sussex. It could only be expected to survive in such conditions as it is not a mountain plant, but it is easily the finest clover I know and very beautiful.

The Cretan iris *I. cretensis* may also be seen in the banks along the road and on the open moorland that stretches back from the road. It is sometimes described as a variety of the stylosa iris but seems to me quite distinct. It makes large tussocks of very narrow leaves. The flowers are only slightly raised above the clumps and are rosy-lilac in colour with prominent white and gold markings. It is a beautiful plant but is undoubtedly very different from the Algerian forms of *Iris unguicularis*, the stylosa iris. It is unlike the narrow-leaved forms of this from the Hill of Chronos and the area around Olympia in the Peloponnese which have sometimes been referred to as *I. cretensis*. All those that I have seen from Greece are true forms of the stylosa. Unfortunately *Iris cretensis* does not share the easy-growing way of the *stylosa* and is rarely a success in English gardens. In fact it very rarely flowers there at all.

This area is a good one for the ground orchids and I have rarely seen so many different kinds in a small space, particularly among the *Ophrys*, the bee orchid and its relations. They are extremely variable, in fact not even two individuals within the same species seem to be quite identical and even to put them into their species with certainty is difficult. Nevertheless they are infinitely exciting and beautiful in their form and markings. Crete has its own species of *Ophrys* in *O. cretica*, a tall stout plant with pink and green lateral sepals which are broad and pointed, and a broad deep maroon lip with a horse-shoe shaped marking. In this it resembles *O. sphegodes mammosa* which is better known and slightly larger in flower, but in *O. cretica* there are usually more flowers and it is a taller plant. All these species are pleasantly and slightly fantastically insect-like. *Ophrys scolopax*, which is generally represented in Crete by its sub-species *cornuta*, which has a pink upper erect sepal with two prominent pink side lobes and a deep brown lip marked like an elaborate bumble bee, is known as the Woodcock Orchid. All the sub-species of *O. scolopax* can usually be distinguished by the little protuberance like an incipient tail that has never developed, at the base of the lip, while those of *Ophrys fusca*, which are also common, can be recognized by the greenish lateral petals. The finest of these that we saw in Crete was the sub-species *iricolor*, which has broad mirror-like blue markings on a dark lobed lip which is extended downwards. It is a lovely and unusual plant as is also Venus's Mirror, *Ophrys speculum*, which has an even larger reflecting blue patch in the centre of its lip. This is bordered by a yellow margin while the outer edge of the lip is very hairy so that it is one of the few species which are always easily recognizable. The yellow *Ophrys lutea*, with a wide yellow border to its lip and rather smaller flowers, is also common and easy to recognize. Among the species of *Orchis* the commonest are *O. italica*, a stout plant with pink and white flowers and long tails to the lip, and *O. simia* which has deeper pink tips to the tails of the lip. The whole effect is like a cluster of acrobats dancing with legs and arms widespread. I doubt whether monkeys ever dance quite like this. The butterfly orchis, *O. papilionacea*, is another which is quite distinct with its very wide flat lip, spread out horizontally. In Crete I have found it very variable in colour and have seen some really exceptional forms.

All these are well worth stopping to photograph but they should not be picked or dug up. Although they produce seed very freely the seeds are minute and have many hazards and the seedlings take from ten to twenty years to reach flowering size. They also have to combine with a fungus to form the mycorrhiza which lives symbiotically with them in their roots and helps them to absorb mineral nutrients. This is present where other orchids are growing but very often not in our garden soils at home, nor in the sterilized potting composts many of us use now. Orchids are particularly vulnerable as species, and they are so lovely in their native habitats that every effort should be made to protect them there.

The bright pinkish-purple *Gladiolus segetum* is unusually fine in Crete and nowhere else have I seen such magnificent patches of it, colouring the landscape like the chrysanthemums or the Viper's Bugloss. Its colour just escapes being magenta. The flowers are more widely spaced than in the common gladioli of gardens and so the whole effect is more graceful. They are particularly abundant in the area south of Phaestos between the temple and the sea, and inland from the caves where the hippies dwell. In England they are somewhat tender and *G. byzantinus* from Turkey and Asia Minor seems to naturalize more easily but has more magenta and not quite so much rosiness in its colour.

Another plant that is particularly fine in Crete is the great Dragon Arum, *Dracunculus vulgaris*, although it is found also in Greece and I have seen gigantic specimens from the west coast of Turkey. It really justifies all the adjectives of fantastic and monstrous. The spathe makes a great shield, sometimes a foot and a half tall and a foot wide, of deepest reddish-maroon while the spadix is longer, a great shining almost black club in the middle. The whole stands on a mottled fleshy stem of a foot or more, and grows out of a rather swollen bulbous green base surrounded by the much divided leaflets of its big almost fern-like leaves whose bases are also mottled with deeper green. I have seen specimens four feet or more tall, but the usual is two to three feet. They grow at the edge of woodlands and in rather dampish meadows in the lower areas, and no other plant is likely to be confused with them. As their colour indicates, they are fly pollinated and as the flower dies it certainly does begin to smell, but there is much less odour to be noticed in young flowers. They are not, however, plants for picking. *Dracunculus* is easily distinguished from *Arum* by its divided leaves. The greenish-yellow *Arum italicum*, like a larger plant of our common Lords and Ladies, is abundant also in Crete, and we have found there also the much smaller Friar's Cowl, *Arisarum vulgare*, with its striped hooded flowers of maroon and green – a perky little flower much more upstanding than the more commonly grown Mouse Plant *A. proboscideum*, and looking not unlike a miniature *Arisaema*.

Fritillarias are very scarce in Crete and the only one which we have found is *F. messanensis*, a delightful little maroon and green chequered bell. Even that, though, seems to be very rare and occasional and we only saw one specimen growing among the rocks of western Crete the last time we visited the island. It was photographed by nearly thirty people but left intact, but it is clearly a vulnerable species now.

The shrubs mostly are much the same as in Greece and equally floriferous. The golden gorse *Calycotome* is abundant and so is the myrtle. A more unusual plant is the pinkish-purple sweet-smelling *Daphne collina* which we found in one area only, growing among rocks and forming bushes about two feet high and rather more across, covered with flowers, like those of a typical daphne.

Another very unusual plant which we found on our last visit, thanks to the interest of a friend who had visited it, is the dwarf form of *Phoenix dactylifera*, the date palm known as sub-species *theophrasti*. This was only recognized as a separate form and described recently by Mr W. Grenter, a noted Swiss botanist who has made a particular study of the island. It only makes a short stem as opposed to the usual long one, but the fronds are six feet or so in length, making a small tree or bush looking like a large cycad and just about as prickly and tough round the base of the fronds, as I found out when I collected part of a leaf as a botanical specimen and a little of one dried flower head. It is quite abundant right on the sea-shore of the north-east around Vai, where there was also a colony of hippies living, very happily it seemed, in tents by an idyllic beach under the palms. We found it again near Canea and there is also a location for it on the south coast, and probably more if one looked persistently for it. It is very fitting that it should be named after Theophrastus, one of the early Greek herbalists whose knowledge of the native plants was very considerable. This is the first herbal in which the author really seemed to know his actual plants rather than merely repeating old legends about them.

RHODES

This is a splendid island for flowers. The flora is closer to that of the western Turkish coast but again it has its own particular and rather exciting endemics, while I have also seen there some of the most splendid groups of ground orchids. There is great variation also from limestone crevices among the rocks to the sand and acid-loving flora around the site of the ancient city of Kamiros, which must have been vast, while the mountains have their own particular endemics. It is surprising how quickly the flora changes with the basic rocks, from the heathers and arbutus and dwarf shrubby *Hypericum ericoides* to the limestone crags around.

One of the most beautiful drives crosses the island from Rhodes to the ancient citadel of Lindos, passing through cultivated land and cistus scrub. One climbs up to the citadel by a steep stone staircase winding through the rocks from the little village below, and the ascent is very rewarding in April for its flowers. There is the brilliant scarlet *Ranunculus asiaticus* in masses which in Rhodes takes over from the white form of Crete and the yellow one of Cyprus and is probably the finest of all, great glowing satiny deep scarlet globes with black centres, a foot and a half tall. This is the form that stretches intermittently right down the eastern Mediterranean coast to Lebanon and Jordan. It looks particularly well against the grey rocks, or indeed against the grey stones of ruined cities as in the great colonnaded market-place of Jerash in Jordan, surely some of the most graceful and dramatic ruins in the Middle East.

In clefts of the rocks has grown the long-belled form of *Campanula rupestris* known as *anchusaeflora*, spreading its beautiful streamers of dark mauvish-purple among the grey stones. It is certainly the finest form I know, the flowers being like a long narrow trumpet, much longer than those of the type. This is the form that should be cultivated, but it is hardly likely to be hardy, since it grows only a little above sea level and here not far from it. Then one emerges on to a platform of rock at the top where are the honey-coloured columns of the old temple perched on the acropolis above the wine-dark sea. It is one of the numerous temples dedicated to Athena, but around it are the remains of battlements built when the Knights of St John were in Rhodes and brought new life to the island. Now the old walls and the pillars rise in April from a sea of cream-coloured chrysanthemums, each with a deeper yellow or orange centre. It is one of the most dramatic sites that I have seen in the Mediterranean.

Another very rewarding drive from Rhodes is up the mountain of Profitas Elias. We go along by the sea first where are splendid clumps of the yellow horned poppy, *Glaucium flavum*, the great yellow globes so well complemented by the blue-grey foliage, and then slowly mount to the hills. The observant traveller who stops at intervals may see plenty of *Ophrys* and other orchids, some indeed of the finest coloured *Ophrys* I have seen, mostly forms of *O. sphegodes*, but all with fascinating although very variable deep brown markings on a swollen purplish-crimson lip. Then there are great clusters of spikes of *Orchis sancta*, a pale chocolate maroon and pinkish-red not unlike the spikes of the bug orchid *O. coriophora*. It is a difficult range of colours to describe and the flowers are larger than those of *O. coriophora*. Here again I have seen some of the finest thistles in the Mediterranean, especially the deep golden *Scolymus hispanicus*, a very decorative plant that is common enough in most eastern Mediterranean countries but seemed very fine on the lower slopes of the mountain here. It also rejoices in the strange name of Spanish Oyster Plant. Even the flower heads are very spiny, while all along the stems are wings with spiny teeth so that there is really no place where one can grasp the plant with impunity. But even greater glories are to be found farther up when one enters the woods. Practically the whole of this area is carpeted with cyclamen. This is *C. repandum* var. *rhodense* and it has rather tall white flowers and a pinkish-purple rim at the base or the throat of the flower. The petals are graceful and twisted as in the more typical *C. repandum* and the leaves are ivy-like. It flowers freely during April and the earlier part of May. With it can easily be found the great pleated leaves of *Colchicum macrophyllum*. These are very large and have the fresh bright green and decorative pleated effect of a veratrum, and when mature will grow up to two feet in height and a foot across, a great shield of a leaf. I have also seen them in the south of Crete and on the Turkish coast very locally, but otherwise it is practically an endemic of Crete. I have never seen the flowers in the wild but it is late autumn-flowering in November-December. It has, however, flowered for us in the bulb frame, the flowers being large and rather egg-shaped, pale lilac and faintly tessellated like the flowers of the commoner Greek *C. sibthorpii* or the rarer northern Greek *C. bowlesianum*. It is well worth growing for its foliage and seems to flower after a good summer baking in the bulb frame.

With the cyclamen also grows the white paeony of Rhodes, *P. rhodia*, and like the cyclamen this seems to be completely endemic to the island. It is a fine species, although perhaps not quite so magnificent as *P. clusii* of Crete. It grows up to two or even two and a half feet with finely divided foliage, although slightly less divided than in *P. clusii*. The flowers are quite large, about three to four inches across with overlapping petals and a boss of golden stamens. Both it and *P. clusii* are slightly tender in cold gardens but it can be grown outside in many, and once established seems to be very long-lived. This, however, is a plant only of the woodland.

These are some of the highlights of the flowers of the island but a longer time spent there and visits at other seasons than spring would surely yield a valuable result.

CYPRUS

The great glory of Cyprus is undoubtedly the masses of *Cyclamen persicum* which decorate the lower slopes of the northern Kyrenia range in early spring. We found them still in plentiful flower in early April but it was a late season and they are recorded as flowering from the beginning of February onwards. They are even found on the promontories of the northern capes which stretch out like fingers into the Mediterranean, and our tables were decorated with great vases of them when we stayed at the St Andreas Monastery in the north-east of the island. There they grow among the scrub which covers the promontory. It is perhaps the most lovely of all cyclamen, with quite large flowers ranging from white with a purplish ring at the base to a beautiful deep shell-pink, about six to nine inches tall, the petals gracefully waved and delicate. It seems far removed from the big hybrids raised and selected from it which florists supply in the winter. The leaves are prominently marbled but very variable both in shape and in colouring and the keen gardener will perhaps select his plants not only for colour and size of flower but also for leaf markings. Unfortunately, they are not hardy outside in England and need to be grown as cool greenhouse plants and repotted at intervals. Some of the tubers are enormous, but it is waste of time to try and collect these since small ones, an inch or two across, establish themselves so much better. The Persian cyclamen is distinguished from all other species by the stalks of the seed heads which do not curl, and I know of no evidence that it has ever crossed with any other species.

On the northern promontories forms of the autumn-flowering *Cyclamen graecum* have also been found, but although in the past forms from here and from Crete have been given individual names the differences are slight and cannot reliably be used to prop up separate specific names. The other autumn-flowering cyclamen of Cyprus is *C. cyprium* which grows more in the southern range and foothills of the island. It is distinguished by the ivy-shaped leaves, deep crimson underneath, and its rather small white spicily scented flowers. We saw it in leaf only in spring in the cedar forest, remnant of the great woods of *Cedrus brevifolia* which are thought to have covered much more of the island at one time.

The cedars in the south-west of the Troodos are now restricted to one small area but it was good to see that they are regenerating freely, while the authorities have planted some of them by the roads in other areas of the massif, so as a species they are not in danger at present. Some of the trees which grow in one sheltered valley are fine specimens, fifty feet or more in height. The needles are shorter than in the Lebanon cedar and a fresh green in colour but the way of growth is much the same. Where they grow it is a cool and slightly moister mossy valley and the trees have established their own microclimate below their branches. As well as the cyclamen there are ornithogalums and colchicums and other small bulbs. The flat spreading growth of the trees can be seen as cedar unmistakably even from a distance.

All three colour forms of *Ranunculus asiaticus*, the Turban ranunculus, are found in Cyprus though the yellow is the predominant form. The whites did not seem quite so large and clear in colour as the whites of Crete but this may have been due to an unusually dry spring. Some were pleasantly tinted pink as also occurs in Crete. The scarlet forms again seemed smaller in flower than those of Rhodes or the Lebanon but we saw them in several places. Still, they show how much Cyprus is a meeting-place for flowers both from east and west. Some of the ones collected in previous years have become established and flowered well in our bulb frame and we hope to spread them more by seed. They are superb and graceful flowers, and to my taste so much more lovely than the double florists' forms.

Mt Troodos is a great massif in the centre of the island and it is composed of basic rocks instead of the limestone of the northern range. The highest point, Mt Olympus – 'Chionistra' on some maps – is also about twice as high being just over 6,000 feet. On it is a great round radar dome, gleaming white like a gigantic egg, an unexpected object to find on such a mountain and a landmark for many miles around, but somehow rather decorative. Among the rocks around it as the snow melts crocuses come up in great masses, *Crocus cyprius* and *Crocus hartmannianus*, only distinguished by the ribbed or annulate coverings of the outer tunics of the corm. They vary in colour from white through pale mauve to quite a deep mauve, always with a yellow base and prominent fleckings on the outside of the flower. In Cyprus they take the place of the *Crocus vernus* of the Alps, and have the same starry delightful appearance although perhaps they are a little smaller. It is an unexpected flower to find on these bare rocky hillsides and they grow mostly among shale and rocks rather than coming up through snow-browned grass as in the Alps.

Near to the top the only other plants in flower under the dark Corsican pines were the charming little yellow *Ranunculus cadmicus* sometimes described as a separate Cyprian sub-species, and *Corydalis rutaefolia*, another dwarf beauty. The buttercup grows up through the screes and pine needles and has fleshy foliage rather like that of a winter aconite, while the flowers, about three to six inches high, are buttercup-like, single on the stems but a good colour. The *Corydalis* only grows three to four inches tall and has glaucous leaves and a short flower spike of several pale pink flowers, each tipped with deeper crimson at the ends of the long spur and also the upper and lower petals. It is like a little orchid superficially. Below ground it has a fleshy round tuber connected with a very narrow stem perhaps a foot long and very difficult to collect intact, in fact it is better left where it is and grown for gardens from seed, albeit a slow process. There are also yellow starlike gageas and white Star of Bethlehem, and in the autumn colchicums of which we were able to see only the leaves in the spring, but in general it is a surprisingly bare and empty flora in the higher zones, although the pines are not thick and do not provide a closed canopy. Possibly goat grazing later, that supreme curse of Mediterranean flowers, is the factor that keeps down the flora. Lower down there are paeonies and pink arabis and large mauve violets and thymes, asphodels, cistus and a big white astragalus, *A. lusitanicus*, known locally as the

'Hairy Milk-vetch'. The only other plant which is abundant in the higher zones is the grey-leaved *Alyssum troodi*. This is close to the better known *A. saxatile* and makes large grey tussocks, flowering yellow in May. There were, however, no signs of flower on it when we visited the mountain in mid-April, but we were told it was an unusually late season. The pines are said to be the Balkan sub-species *pallasiana*.

Arabis purpurea is a very fine species, growing usually in vertical clefts of walls or in scree, usually avoiding very sunny slopes. The flowers are slightly larger and taller than those of our common white arabis. It makes a lovely effect from a distance. In the northern range and among the walls of St Hilarion Castle close to Kyrenia grows *A. cypria* which is very similar in size and colouring, and I feel that a later authority may possibly decide to merge the two species under one name.

The paeonies were not in flower during our visit in mid-April but in large bud. They have been distinguished as *P. arietina* var. *orientalis* in Sir Frederick Stern's *Genus Paeonia*, the varietal difference from the Turkish and Greek forms of the species being based on the smooth leaves, stem and petioles. We saw a large and flourishing colony about a thousand feet below the summit and were glad to notice that it was seeding freely. *Paeonia mascula* is also reported to grow a little lower down, but the plants we saw seemed uniform enough and to be all one species, with the larger number of leaflets of *P. arietina*. They must be a magnificent sight when in flower towards the end of April and early May. In the few damper places where a little stream was still flowing there were lovely big violets, with pale violet-mauve flowers, probably a form of our dog violet *V. canina*.

The thyme was *T. integer* and grew abundantly over the lower slopes of the mountain. We did not see it in the northern range and the floras of the two seem very distinct. It formed large mats, sometimes a foot or more across, covered with pinkish lilac flowers. When one looked carefully one could see how distinct these were, with very long tubes to the flower. It was certainly one of the finest thymes that I have seen and I hope that it may grow reasonably in a warm dry position in southern England.

Astragalus lusitanicus is mainly a native of Spain and Portugal as its name suggests, but in Cyprus the eastern var. *orientalis* was abundant on the lower slopes of Troodos. It is a statuesque plant up to two feet tall with silky grey leaflets in large compound leaves and white pea-like flowers in quite bold racemes, but it misses real distinction since the whiteness of the flowers is not of the purest and has a slightly milky tinge. I have not seen this plant in Greece or other eastern Mediterranean islands so it is strange to find in Cyprus such a predominantly western species.

Among the trees the evergreen golden oak *Q. alnifolia* is probably an

THE MOSS FOREST OF BORNEO

(*Left*) A species of Coelogyne, an interesting orchid growing on a carpet of moss.

(*Right*) In the early morning the mist rose from the forest below in clouds to envelop the moss forest.

THE GIANT PLANTS OF THE EAST AFRICAN EQUATORIAL MOUNTAINS

(*Left*) Giant tree heathers on Ruwenzori

(*Right*) *Lobelia elgonensis*, a stiff green six-foot obelisk on Mt. Elgon in Uganda.

(*Above*) Giant lobelias and senecios near the crater of Mt. Muhavera in S.W. Uganda. The lobelia is *L. wollastonii*.

(*Below*) Stuart Somerville painting among the giant heathers and lobelias of the Nyamgasani lakes on Ruwenzori.

LILIES IN AMERICA

(*Above*) *Lilium canadense* which has lovely yellow flowers in the Eastern States.

(*Below*) *Lilium washingtonianum* var. *minor*, a rare species on Mt. Shasta in the Pacific States of the West Coast.

endemic of the island, plentiful on the slopes of Troodos. It usually makes a large bush rather than a tree, and is dark in the landscape with hard leathery leaves which are only golden on the under surface. The gold is rather an olive-gold than a bright yellow but still it is decorative when seen close up although making no effect at a distance. In April the oaks were covered with long acorns in little encrusted cups.

Arbutus andrachne was particularly fine in flower both on the Kyrenia and the Troodos ranges, great creamy umbels of billowing bells among the fresh yellowish-green leaves, while the cinnamon-red trunks twisting and curving to the light were always a delight. The Judas trees in full flower were as fine as any I had seen, large trees in some of the gardens absolutely deep purplish-red all over, flowering well before the leaves. They seem to need the Mediterranean sun to ripen them well to induce this abundance of flower. All along the banks of the Bosphorus above Istanbul they are equally magnificent in April.

The terrestrial orchids were a great feature of the visit, especially *Orchis anatolica*, a predominantly Turkish species, and *Orchis simia*, the monkey orchid. They were plentiful beside the forest road which runs along the upper slopes of the Kyrenia range, together with various *Ophrys* and *Serapias*. *O. anatolica* was very variable in size and we saw some massive spikes up to eighteen inches high. The flowers are well spaced on the stem, pale blush purplish-pink with long spurs deeper in colour towards their tip and a whiter marking at the centre of the lip dotted with small deeper purple spots. The spur usually points horizontally or upright. The monkey orchid is a more shaggy plant with flowers more closely aggregated on the spike and longer tails to the lip. Here the spur is much shorter and points downward while the tip of the spurs and the extensions of the lip were deep purplish-maroon or in some cases a violet maroon, a feature which I had not seen in other areas. Each floret is really like a little paper toy monkey dancing on a stick. So also are the florets of *Orchis italica* which has a rather larger and even shaggier spike. We did not see quite the same range of orchids that we had seen previously in Crete, but still they were very lovely. We stayed a night at the St Andreas Monastery near the end of the north-east peninsula of the island, and the dining-hall tables (one could hardly call it a refectory), were all covered with orchids from the neighbourhood – *Ophrys sphegodes mammosa* and other *Ophrys* and masses of *Serapias*. Conservationists shuddered at the picking of so many and it will certainly endanger their numbers in the future if the practice is continued.

The same practice had diminished vastly the population of the beautiful deep wine-red *Tulipa cypria*, an endemic of Cyprus from the north-west of the island. We were offered bunches at a local cottage but saw literally only two or three in the fields and one patch on a slope where formerly thousands were recorded growing among the corn. Now they have to be sought higher, out of the range of cultivated land, but if they are picked continuously with their leaves they will become so scarce that in time the practice of gathering them will not be worth while and the numbers may recover a little. Probably deeper ploughing and the use of herbicides have contributed also to the disappearance of this tulip from its old areas in the cornfields. It is a very fine species with large flowers, each petal with a basal blotch, deep maroon-black in the centre and edged with yellow. It is almost the size of a small Darwin tulip and must be one of the largest in flower of the Mediterranean species.

Cyprus also owes much to its imported flowers. The island is very warm and as a street tree there are masses of purplish-pink Bauhineas from India with their big exotic flowers – very orchid-like in appearance. The mimosas have spread freely into great groves in the more sandy places. These are mostly *Acacia cyanophylla* with narrow entire leaves and more orange-yellow flowers than our common mimosa, but they were weighed down with flowerlike fluffy orange balls. They have been planted and now spread freely all round the excavated site of the old city of Salamis. Curiously one does not see the florists' species, *A. dealbata* and *A. baileyana*, with their more graceful and finely divided foliage. The Cassias in the gardens were also magnificent and covered with small orange flowers. As long as there is water Cyprus could be a gardener's paradise, especially in the spring. Among the fields and on wastelands the Bermuda Buttercup *Oxalis pes-caprae*, a native of South Africa, has spread widely as in many other Mediterranean countries and is a very lovely flower with its clear lemon-yellow cups. The annual chrysanthemums were magnificent in early April and gave bright colour with sheets of golden-yellow. We also saw on the central plain sheets of pale blue flax in full flower contrasting with the much deeper blue anchusa and the magenta-pink *Gladiolus segetum*, a warm and strong colour and a very beautiful flower, so much more delicate and graceful than our massive large cultivated gladioli. Even more exciting for me, however, was the little dwarf gladiolus of the south, *G. triphyllus*, which has flower spikes only six to eight inches high with brighter pink and white flowers. The three upper petals are pink and the lower three, which are narrower, are a creamy-white. It gives an orchidaceous bicolour effect and is unlike any other species of gladiolus that I know, as well as being the smallest.

The cistus of Cyprus are also magnificent, all shades of pink and magenta of the vast aggregate species now known as *Cistus creticus*. In the north, especially along the promontories, the rather small-flowered forms are a deep clear pink, as pink as any I have seen; they are a clearer pink than the flowers of *C. purpureus* and much lighter and without a blotch. Probably they belong to the group often separated out under the name *C. parviflorus*. Under them on these windswept headlands grows a particularly bright blue dwarf form of *Alkanna tinctoria* and also a bright blue *Lithospermum*. The common *Lithospermum* of the island, however, is the prickly shrublet *L. hispidulum* with small tubular blue flowers which turn pink as they age. It is a fine plant as seen in the wild although I doubt whether it would make a good garden plant.

CHAPTER FIVE

Eastern Turkey and Persia

EASTERN TURKEY

The expedition here was only a modest one of two months but we covered a great distance and it was one of my most interesting and satisfying experiences. Admiral Paul Furse, who had just retired from the Navy, was my colleague and his vast knowledge of fritillaries gave us a good basis of areas to visit. The Council of the Royal Horticultural Society for whom I was working very kindly gave me two months' leave after fifteen years' service as editor of their publications, to visit two countries that we had both long wanted to see, the high plateau and mountains of north-eastern Turkey, the great Pontic ranges which stretch inland from the Black Sea coast and the mountains of northern Persia to the south of the Caspian. These are great areas for bulbous plants of many kinds and also for the strange and beautiful Oncocyclus irises which had long fascinated me.

For fifteen hundred miles there is hardly any ground below 4,000 feet and much of it is over 8,000 feet with a number of high mountain road passes. These were invaluable to us, enabling us to get quickly and relatively easily into the higher zones. These areas are particularly rich in bulbous plants – tulips, irises, crocuses, colchicums, sternbergias – while Asia Minor is probably the centre of the Asiatic group of fritillaries in which my colleague, Paul Furse, had long specialized. I expect that many have seen his drawings in the R.H.S. halls, but he didn't unfortunately have much time to draw on this trip, since we covered just over 6,000 miles from Ankara and back to Ankara again in the two months. Persia alone is three times the size of France, and Turkey is nearly as big. We could only regard this trip in the nature of a reconnaissance for more thorough searches of particular areas which might be attempted later.

So we had to use modern means of air and car travel and then it was possible to cover quite a large area. The aeroplane has made great changes in one's routine. Not only can one get near the collecting ground quickly, but one can also send back plants in polythene bags by air freight; we sent three such parcels from Tehran which arrived at Wisley within two or three days of the time of sending.

84

Just after the Rhododendron Show in early May 1960, I flew out to Istanbul by a Comet plane. In less than eight hours from London one can be looking at the magnificent mosques and minarets of Istanbul, surely some of the finest in the world. St Sophia, which was built as a Christian basilica in the year 532 by the Emperor Justinian, enriched and partly rebuilt by later emperors, made into a mosque in 1453 after the conquest of the city by the Turks and secularized as a museum by Ataturk in 1935, remains for me as one of the marvels of the world and the odour of sanctity is with it still. The great dome, one of the largest in the world, larger than that of St Paul's, seems to float in the air over its square base; the proportions of the architecture are completely right and make it infinitely satisfying. A thousand years later the great Turkish architect Sinan built mosques nearly of an equal quality all over the city. The finest and most impressive is probably the Suleymaniye Mosque with its grey domes building up to a crescendo in the great central dome between two half-domes. He can be ranked as the Christopher Wren of Turkey and must have built nearly as many mosques as Wren built churches. The other great church of the city, the Kariye Charme, which all visitors should see, is only a small one and built in the early years of the 14th century but its walls are covered with the most exquisite mosaics of scenes from the life of the Virgin Mary and Christ. Recently there has been some very skilled restoration and they gleam with gold and jewels as they did originally indeed a rich mosaic of colour surrounding long rather sad faces. Still, it is a masterpiece.

But one of the great thrills of Istanbul is the teeming life all around, one's first taste of Eastern life and European life mingled. The waterway of the Golden Horn divides the city, and it was a treat to breakfast on the balcony of one of the hotel's rooms overlooking the waterway and see the shipping jostling along in front. Alas, I have never had sufficient time in Istanbul really to get to know the city and savour it fully or explore its surroundings, although I have been there several times.

A day later I went on to Ankara where I found Furse who had driven out more leisurely with his wife in a Land Rover, specially obtained for the trip. We were also joined there by Mr Derek Hill, the artist, who travelled with us to Tehran and from his enthusiasm for Seljuk and other Moslem architecture added much to the interest of the journey. He has since published a great book of pictures of Moslem decoration. The Seljuks were one of the great dynasties of Turkish rulers and left very fine buildings behind them. While Derek travelled with us, wherever there were old sites near the route we would spend the morning botanizing but sometimes deviate in the afternoon after 'Seljukery' and wonderful old caravanserais, so much finer than the Western caravan park yet with some similar amenities. The caravan traveller then coming perhaps from 'The Golden Road to Samarkand', needed accommodation at night, and the buildings round an open courtyard would be split up into separate bays for each family or group. At other times the quest would result in an ancient domed tomb, covered with decorated tiles. Ankara was not nearly such an interesting city as Istanbul and we were glad to move on as quickly as we could. The country round is mostly open steppe and we found there a few crocuses, probably *C. ancyrensis* with bright golden-yellow flowers, and greenish-yellow dwarf irises, probably *I. schachtii*, a little known species, also some flowering cherries and plums but little of great interest.

'The Rose of Persia', as the Furses had christened our vehicle, was a Land Rover with a long wheel base and extra creep gears and four-wheel drive and other refinements which made it almost ideal for the job. It took three in front, could be locked up, and held a vast amount of stuff inside. This included several cubic feet of paper from Kew, weighing several hundredweight, for the pressed herbarium material, a good supply of trowels, picks, etc., and polythene bags for the live bulbs and plants. It is very difficult for a botanist and plant collector to travel really light and we used all the presses and drying paper. We were also completely self-contained for camping under rather varying conditions. Even in mid-May there was quite a lot of snow on the Turkish and Persian mountains, nor was it all gone when we came back in early July, while in many other parts of Turkey and Persia it was very hot by day, though nearly always adequately cool at night.

Paul had displayed a naval facility for expert packing and everything was neatly in its place. So expertly lashed in fact were the camp chairs and tables on the roof that we never unpacked them the whole time, and did our specimens of an evening on the tail-board of the Rose which let down to form quite an adequate table. We had one tent which we used sometimes, especially in the early part of the trip although I tended to use it more than he did. Often in Persia when it was hot we both slept happily outside.

From Ankara we crossed the Anatolian plateau aiming north-east for Trebizond. By three separate routes we climbed over the Pontic range which runs for about three hundred miles along the south of the Black Sea, ranging from 7,000 to 8,000 feet in the western end to nearly 13,000 feet in the eastern end near the Russian border. It is from here that so many species called *pontica, colchica* or *iberica* have come. These are nothing to do with Spain and many people tend to be easily misled by the epithet *iberica.* Probably our best plant on the Anatolian plateau near Corum was a most attractive small prostrate shrubby convolvulus named later as *C. assyriacus.* It had silvery foliage in small domes and large pink sessile flowers. Unfortunately there was no seed and we did not manage to keep the plant during the long drive to Tehran. It would, however, be well worth collecting if one should pass that way again. It grew among pavement-like rocks not far from the road, but we only saw it in one area. Every expedition has its regrets for the plants it would have liked to have introduced to gardens and failed to do so. Another in this case was an exciting *Aristolochia, A. pontica,* with stems up to 2 feet, heart-shaped leaves, quite large flowers several inches across of creamy-yellow, brown and green and shaped like a Dutchman's pipe with a pale green parrot's beak hanging down. It grew in beech woodland near the top of the Pontic range in the western end and below ground it had a large swollen tuber like a mis-shapen elephant's foot.

Our first crossing of the range was from Niksar, a small town at the edge of the plateau northwards over the range to Unyie on the Black Sea coast, and this was a rewarding trip on which I wished we had had more time but we still had thousands of miles to go. It had rained and the road up was so wet and the soil so heavy a clay that we needed to use our lowest gears, while the next morning we found ourselves in a cloud. We camped in a very fine beechwood and during the night heard wolves howling in the woods but none molested us and we could not tell how near they had come to the camp. Very early in the morning as dawn was breaking we heard a thumping on the tent and a gruff voice spoke in Turkish which unfortunately we could not understand, and after a few minutes of futile talk and gesture the two Turks left us. Probably they only wanted a lift down to the coast, several hours' drive below, but we felt alarmed and after a quick breakfast we packed up the camp and drove off, at this early date perhaps not so confident of the general friendliness as we became later.

Near the top of the pass on the north side were thickets of the lovely yellow azalea *Rhododendron luteum,* growing up to six feet. The flowers were as strongly scented as those of the English gardens and the shrubs were just as floriferous and vigorous. There was some variation in depth of colour and size of flower head. The range of this azalea was from the top of the pass on open moorland down to the upper limits of the beech forest where some very fine trees were seen. Later in the day, however, we found it right down by the sea-shore. It was too grey and misty to savour its pungent scent or to get good photographs. Much of the undergrowth consisted of a dwarf shrubby *Gaultheria* with small pink flowers in clusters which I thought might have made quite a good horticultural plant although it might have been over rampant, as is *Gaultheria shallon*. No one will thank you for bringing home such a menace and it is difficult to tell in the wild. There was also the very sweet-smelling *Daphne pontica* with its acid yellowish-green flowers. It grew at the edge of the beechwoods and makes a good garden plant although the flowers are rather hidden by its thick evergreen foliage. A little lower but overlapping the yellow azalea was an old friend, or to some an old enemy, *Rhododendron ponticum*; but clothing the grey granite rocks and cliffs of these mountains in great masses in full flower it was very fine, just like a bit of Scotland, while one member of the party described it as being 'like driving up to Exbury', perhaps rather a libel on this famous rhododendron garden near the south coast of England. It was interesting to notice that it was both as vigorous and as variable in flower colour as it is in this country and from this I would tend to doubt theories which suggest that all our plants are of hybrid origin.

As we neared the coast, in the undergrowth – mostly of hazel – were masses of cyclamen leaves but no flowers. These were rather small, rounded and marbled on the upper surface. We collected some and they have flowered at Wisley, mostly good pink and purplish-pink forms of the very variable aggregate that is now known again as *C. coum* but has also been known as *C. vernum*, *C. ibericum* and *C. orbiculatum*. Near the coast were tea plantations, a species of *Camellia*, but rather dwarf here and probably at the northernmost limit of its range. Whether the ubiquitous tea in little glasses came from local stocks or from imported leaves we did not find out. The Black Sea coast was beautiful and I wished we could have lingered longer but it was already well after dark when we reached Trebizond. Having recently read Rose Macaulay's fascinating book *The Towers of Trebizond*, I was expecting a romantic Eastern city, but instead we found a large commercial town with much traffic, and oil tanks and installations near the sea. The more interesting buildings and old Byzantine walls were now rather submerged in the sprawling growth, but the old basilica of St Sophia still stood on a green hill just outside the town and the wall paintings were being patiently uncovered by a party from Scotland and were very fine. They were, however, under a constant threat that the Turkish authorities might decide to whitewash them over again since the basilica was now also a mosque, but I believe that this has not taken place. Trebizond had been the last seat of the Roman emperors before the Turks captured the city in 1461. We had, however, to wait until we got to Erzerum to savour the real Middle Eastern atmosphere.

From Trebizond we drove eastwards along the coast although we did not have time to go as far as the Russian frontier. From Rhizeh, a name well represented in botanical literature, we turned inland in to the mountains along a small road hoping to find that curious ericaceous plant with pink cup-shaped flowers, *Orphanidesia gaultherioides*. The usual form is pink but I felt that there might be just a possibility of finding a white form; but, we did not see any sign of it and after about twenty-five miles the road had been washed down into the river and we were unable to take the car further and did not have time to go far on foot. It could, however, be a rewarding area. Here also had been collected *Galanthus rhizehensis*, still a rare snowdrop. This northern side of the range is an area with a very heavy rainfall and lush growth and a great contrast to the southern side. However, we did collect several interesting terrestrial orchids including some lovely *Ophrys*, *Serapias*, *Orchis* and *Habenaria* species some of which have grown at Wisley and the flowers were identified at Kew. However, even in an area as remote as this one should be sparing in the orchids one collects. Turning back along the coast through Trebizond we drove up the famous Zigana Pass, a magnificent road in every way. It was down this pass that Xenophon had led his army to the sea and there they had been stupefied from eating yellow azalea honey.

The beechwoods and azalea thickets have now been much thinned, but a few large trees remain and at the edge of one grove at about 5,000 feet we found young lily spikes of the *monadelphum-szovitsianum* group. Some of these were in flower on our return journey at the beginning of July; they were *L. szovitsianum* and had magnificent spikes of yellow up to five feet, and again they were variable both in depth of colour and spotting as in English gardens. Growing with them were *Veratrum* and *Aquilegia* and *Galanthus*, all without flower. The aquilegia later turned out to be the lovely *A. glandulosa*, with large blue and white delicately poised flowers like those of *A. alpina* only slightly larger, and the snowdrop was *G. ikariae* subsp. *latifolius*, growing unusually lush in a rocky pine wood. An interesting cyclamen was also found here with dark rounded unmarbled leaves, but unfortunately without flowers, and we wondered whether we had at last found the habitat of the original unmarbeled *C. coum* which does not seem to be native to the Isle of Cos and whose origin has been something of a mystery since the time of Clusius. These, however, flowered at Wisley and were not very distinct from the forms collected lower down. Higher on the moorland zone near the top of the pass at about 8,000 feet we found another cyclamen with minute round leaves, mostly smaller than a farthing. This was *C. parviflorum*, a species close to *C. coum* and previously known mostly from Russia. The flowers are almost prostrate, rather squat and a deep or pale lilac or almost mauvy-pink, but unfortunately it is not a strong grower. The leaves also are unmarbled. Another interesting plant here was *Anemone blanda* var. *scythinica* with flowers a pale sapphire blue on the outside and white on the inside. It is generally rather less vigorous than the blue forms of this anemone but it is a delightful plant rarely seen in gardens. This was a cold bleak area near the top of the pass, low moorland with a Scotch mist and a little snow in patches. Probably it remains moist throughout the year with snow lying all the winter.

On the slopes the whole pass was full of leaves of a large colchicum, probably *C. speciosum*, and must be a very fine sight when these are in flower in the autumn. Crocuses were scarce but on the grassy moorland at the top of the pass we found a small mauve-flowered species, probably *C. aerius*, and also one in leaf which, when it flowered at Wisley in the autumn, turned out to be *C. vallicola*, a plant very rarely seen in cultivation. It had medium-sized white flowers with two golden spots at the base of each segment on the inside of the flower. This shows how worth while it is to collect such genera as crocus when one sees them even though they are not in flower. We particularly hoped for the golden-yellow, summer flowering *C. scharojani*, but so far it has not appeared among the corms collected and we were not there at its flowering season which is August and early September. It is reputed to be a plant of moist places. It is surprising how quickly the eye gets attuned to the silvery midrib of the crocus leaf and picks them out from the grass around. Later collectors in the autumn have brought back more crocuses from this high area of the Pontus including some golden-yellow ones with pointed petals, flowering in September. This has been identified as *Crocus lazicus*, and probably *C. scharojani* is either a form of this or synonymous with it; but one would need specimens collected from the southern Caucasus, where it grows, for comparison to unravel the difficulties.

From the crest the pass descends to Gumusane, the base for various collecting parties in the past, and then ascends again on the way to Bayburt and Erzerum, and there was a side pass up which we also climbed. Here we found our first paeony, tulip, iris and fritillaries. The paeony was in full flower, large pale crimson globes with great bosses of golden stamens in the middle and was probably a form of *P. arietina*. On our way back we were able to collect seed. The tulip was a brilliant scarlet-flowered plant with a deep maroon base surrounded by a bright yellow band. It has been provisionally identified as *T. armena*. The earlier findings were about eight to ten inches in height but higher up it became almost sessile with stems reduced to an inch or two, but with flowers of the same size. The iris was a dwarf Juno iris which had already finished flowering, probably *I. caucasica*, a plant which we found several times later with quite large pale lemon-yellow flowers about four to six inches high and several flowers to a stem. Although very rarely seen in cultivation, it has quite a wide distribution from the Caucasus through much of north-east Turkey and north-west Persia, always growing on sloping ground, generally surfaced with fine scree. It is important in collecting these Juno irises to keep intact the thick fleshy roots which grow down below the bulb.

The fritillaries were very local and usually more difficult to find. One of the two here was either *F. caucasica* or *F. armena* with narrow bell-like flowers, deep blackish-maroon and covered with a faint blue bloom, while inside they were deep yellow-green. Unfortunately it only seemed to survive in scrubby thickets among the dried-up leaves, where presumably the goats were unable to reach it. Another dwarf fritillary from the same area, with quite large dark maroon slightly chequered bells, was *F. latifolia* var. *nobilis*, which was usually found in fine open sloping scree. Both have flowered at Wisley.

On our return journey in the first days of July we were able to find some of these plants again, in seed, and also to make a crossing of the Pontic range between Bayburt down to Of on the Black Sea coast, and this proved a most successful trip. We reached about 10,000 feet over a pass with open alpine grassland and then descended very steeply to the sea level. In the open turfland we collected plants of two gentians, one of which was in flower and was *G. pyrenaica*. It is curious that, apart from a station in Hungary, this species apparently has no intermediate habitat between the eastern part of this Pontic range and the Pyrenees. It was a good deep blue form growing in the short grass in rather dampish places. The other species was one of the *lagodechiana-septemfida* group. Another interesting plant of the same area, which was quite common, was a dwarf campanula, *C. tridentata* var. *stenophylla*, with large upright to horizontal mauve flowers, white at the base. It was close to *C. aucheri*. The creamy-yellow dwarf *Daphne glomerata* was also found among the rocks and in the short turf, a woody stocky plant never more than eight inches in height, with quite large flower clusters but without any very strong scent. It is a native of the Caucasus and has been a very rare plant in cultivation for some time. In this short turf we also collected numerous crocus corms in leaf, mostly *C. vallicola*, and also a very dwarf cyclamen with small rounded, only slightly marbled leaves, smaller than a farthing. It was probably the same as the small one collected on the Zigana pass *C. parviflorum*. A good deep pink *Asperula*, *A. pontica*, grew among the rocks and was deeper in colour than any I know in cultivation.

This was one of the richest areas we explored and a longer visit would have been worth while. On the northern side of the pass, however, growing in a rather damp, lush, but steeply sloping meadow with a patch of melting snow only just above, we found one of the most interesting plants of this expedition: a tall and handsome Turk's-cap lily up to four feet in height with a loose head of citron-yellow or deep butter-yellow flowers, deep maroon in the centre. The leaves were markedly ciliate in some cases and it had a very large bulb and no stem roots. Unfortunately the scent was reminiscent of *L. pyrenaicum*. This was the plant which was subsequently named *L. ciliatum*, 'Hairy Esau' as Paul Furse christened it, and he found it again on subsequent expeditions and it appeared to be rather variable. We found out also that the plant had been collected previously by Mr E. K. Balls and Dr W. Balfour Gourlay during a collecting trip in the Pontus range many years previously, and a bulb of it was still growing at Quarry Wood, Newbury, that great lily garden started by the late Mr Walter Bentley and now carried on by Mrs Martyn Simmons, who is equally keen. So our ambitions to discover a new lily were disappointed, but still our collecting did result in the plant being named and separated from *L. ponticum*. Just as the light was falling and we were still zig-zagging down one of the steepest roads we encountered on the whole trip although now 1,500 feet below the other lily, I saw a flash of deep yellow among the rocks. Paul was driving, and at first was reluctant to stop to investigate. However, we did stop and it was a lily, the true *Lilium ponticum*. It did not usually grow more than one and a half or two feet tall, with only one or two flowers to a stem, and was a much slenderer smaller plant than Esau. It also had a much smaller stem-rooting bulb and the flowers were a slightly deeper tawny-gold with a large black centre. Again, Paul found this later in other areas on subsequent trips but it was an exciting day on which we found two lilies. Growing with the taller lily were such garden plants as *Geranium psilostemon* with its strong magenta and black flowers each with a black spot at the base, also a tall *Veratrum*, probably *V. viride*. The soil was heavy moist loam and in nearly all cases it was notable that the bulbs we collected grew in the better and heavier soils although generally they were well drained. A patch of pale blue scillas, probably *S. sibirica* var. *taurica*, was actually growing in running water on a steep slope and digging for them was like plunging into damp dough. I have little doubt that this Pontic range contains a high number of potentially good garden plants and will repay further and more detailed searches. We also found high up on the moorland of this pass a dwarf creamy-yellow rhododendron, probably a form of *R. caucasicum*, one of the basic parents of the stocks of many of our old hardy hybrids, but this was a much dwarfer form than usual, only about six inches tall.

The Anatolian plateau route between Bayburt, Erzerum and the

Persian border leads over two quite high mountain passes, those on the flanks of the Kop Dag and the Sarikoma Dag, and both were interesting and profitable areas. In the first half of May the snow was only just melting not far above the road, while in July the meadows were lush and flowery, in fact as full of colour as many alpine meadows just before they are cut.

Particularly beautiful and interesting were the Oncocyclus irises of the plateau and also some of the ones found later in Persia. Between Tercan and Erzerum we found our first one, a tall vigorous plant with large flowers, heavily veined with purple on a pale yellow base, probably *I. sari*, but it seemed a more attractive plant, and with a larger flower than the one illustrated in Dykes' *Genus Iris*. Undoubtedly, however, it was very variable. To the east of Erzerum we found *Iris iberica*, a superb and exciting plant, growing quite plentifully among the short grass of one area. The flowers are very large, and conspicuous from the contrast of white standards and dark chocolate-maroon falls, and are borne on quite short stems often under a foot in height. In the low evening sunlight shimmering through the white standards they were particularly beautiful. These plants were in flower in mid-May and later we found them again in seed in early July but they were very much more difficult to distinguish then. In each case they were growing in quite heavy clay in full sun.

In all cases we found that each species of these Oncocyclus irises was restricted to quite a small area, generally under a mile long, and we never saw the same one in flower in two different areas. These Turkish and Persian representatives of the group seem to have much narrower leaves than the Palestine group and the leaves curve backwards in a bow. In their own habitat they undoubtedly withstand long cold winters with much frost and so one hopes that in this country they may prove a little hardier and more adaptable than the Syrian-Palestine group, and perhaps start to grow a little later. Undoubtedly though, a thorough summer ripening will be required, combined with ample feeding during the growing season. In my experience they grow better planted out in a raised frame rather than kept in pots, as also do crocus, reticulate irises and many other genera, especially if the watering can be from an underground perforated pipe. *Iris sari* has grown quite well in the bulb frame and flowers practically every year, but *Iris iberica* has proved more difficult and so has its Russian counterpart *Iris elegantissima*. Some growers, however, have managed to bring it into flower and to increase its growth. It is certainly one of the finest in flower. Subsequent collectings produced forms with the white standards heavily veined with deep maroon, but these are not so conspicuous.

Erzerum was the first city where we felt we were really getting into central Asia and it was very different from Trebizond. Cradled in a valley surrounded by mountains, fiendishly cold in winter and hot in late summer, it was one of the chief cities of the Seljuk dynasty in the twelfth century. They liked strong colour and inlaid bright turquoise blue tiles into little bricks in their buildings. There was a fine caravan-serai, now a barracks with a large courtyard and a chimney with most elaborate patterns of blue and dark red. Then there was a holy school with a large mosque which must have been a kind of monastery, entered through a gateway with two magnificent tiled and brick minarets. Erzerum was then a military area and one picked up a soldier guard outside the town who escorted one into the town to register with the military police. These were friendly and gave us little bowls of tea, and thence to our hotel. Next afternoon another guard escorted us out of the town. I gather, though, that now this formality is no longer required and most restrictions have been lifted. One can even go without hindrance down to the fabulous Lake of Van with its little ruined early Armenian churches. It was a closed area in 1960 and it was most tantalizing to pass by the turning a little beyond Erzerum with a signpost marked Van. However, other parties have been able to go since then, and have brought back different irises including the Oncocyclus species *I. paradoxa*, with rather reduced falls and a strong purple stripe across the velvety dark base. They have also found interesting fritillaries.

On the mountains we were able to collect further tulips, colchicums and fritillaries, the former only just emerging from the ground as we went through and past flowering as we returned; one tulip was very stout and may belong to the *eichleri* group, while others were much slenderer with gobular capsules.

The *Colchicum*, probably *C. bifolium*, was a pale rosy-mauve, a starlike flower close to the ground which sprinkled damp patches of turf just where the snow had melted, rather as the soldanellas do in the Alps. Unfortunately at this stage no new corm had been built up and the old one was largely used up, so it is likely to take a year or two for the corms to reach flowering size again. Obviously these spring-flowering colchicums only grow in places where they have considerable moisture during their short growing and flowering season. These meadows were still quite moist in early June when we collected further corms and the soil was a heavy, sticky clay like damp dough. However, it was reported to be an unusually late season.

Some irises of the Reticulata group were found, mostly out of flower, but a few *I. reticulata* var. *krelagei* were seen and others were very early flowering dwarf reddish-purple. Some, that have flowered at Wisley, have an unusual bicolor effect which I had not seen before in a Reticulata iris and are an interesting addition to our range. *Puschkinia scilloides*, like a large and paler scilla, was common in damp places, while dwarf ornithogalums and gageas were abundant where the snow had recently melted, but cannot be claimed to have great horticultural value. Three primulas were seen, one a large oxslip like the Swiss one, another larger and with deeper yellow flowers, probably *P. elatior* var. *columnae*, third a rose-mauve primula of the *farinosa* group, probably *P. auriculata*, which grew only in damp places or beside small streams. It is rather taller and stouter than *P. farinosa* and seems to replace it in the ecology of these mountains. It was featured in one of the very early botanical magazines as *P. longifolia*.

An interesting sight was a meadow near Mt Ararat mauve with nodding heads of gladioli, probably *G. atroviolaceus* and *G. imbricatus* for the most part, but including also *G. segetum*, a more rosy-purple. There appeared also to be a number of intermediate hybrids between them. *G. atroviolaceus* is a very graceful and beautiful plant and would seem worthy of wider cultivation if it should prove sufficiently hardy but even now it has not been tested enough. The mountain was very lovely, almost a perfect cone, snowcapped for the last few thousand feet. It was another temptation to deviate but few good plants have been reported there and it is a long climb even up to the snowline. Several parties have climbed the mountain in the forlorn quest for remains of Noah's ark, but it is hardly to be expected that anything tangible would be found that could be identified with certainty as belonging to the ark. Sticking up so steeply from the surrounding lowland, there is no cause to doubt it is a possible landing place. Beside it is Little Mt Ararat, another perfect cone, like a very small brother beside big brother.

PERSIA

The Iranian frontier at Bazorgan did not bring any great change in flora. Already we had entered Azerbaijan, a rolling, hilly country of cornfields and short grass, rocky mountains and fine horsemen. Between the frontier and Tabriz, however, we were able to collect two more Oncocyclus irises, one possibly *I. lycotis*, previously known only from Russia, was particularly fine with heavily veined mauvish-chocolate falls and standards and a large deep maroon blotch on the falls. The flowers were nearly as large as those of *I. iberica* and carried on short stalks mostly under a foot in height. One variant with more purple colouring in both fall and standard was found in this group. It appeared to resemble the Russian *I. elegantissima*. The other species was both smaller and slenderer in growth, a rich mauvish-purple in both standards and falls. This may be *I. polakii*.

The euphorbias of all this area were very decorative with reddish heads and very glaucous foliage, while we also found on the volcanic rocks various pink *Aethionemas*, yellow *Onosmas* and several *Alliums*, one dwarf with large leaves resembling *A. karataviense*, but somewhat smaller and probably *A. akaka*.

Tabriz was our next main stop, a more Eastern and Asiatic city than we had yet visited, dominated by a tremendous fort-like citadel. Here we were particularly indebted to Mr Harold Popplestone the representative of the British Council, and his wife, who introduced us to the Governor of the District and other local celebrities and also took us up the mountains. On the rather dry and bare red hills north of the city we found the remains of Juno irises which had already finished flowering, probably either *I. caucasica* again or maybe *Iris persica*. Unfortunately we were too late throughout Persia to see *Iris persica* in flower.

Of the two big mountain groups near Tabriz we visited the more southern one, Kuh-i-Sahand, which lies to the east of the big Lake Rezaiah, formerly known as Lake Urmiyah. There was still quite an expanse of snow on the range. Our two days on the range were only enough, however, to scratch the surface of it and unfortunately we were not able to return as we had hoped, to visit it again or to visit the more northern group, the Kuh-i-Savalon. Between 7,000 and 8,000 feet the grassland on the eastern slopes gives place to rock and scree with more occasional patches of spiny astragalus and other thorny bushes and some short grass. Here we found an attractive tulip with rather long slender flowers of a rich purplish-crimson, probably a form of *T. violacea*, but taller and not so dumpy as that species usually is, at any rate in cultivation. It was also unusual in having only a flush of blue at the base, without any yellow. Other forms collected later on the other side of the lake have flowered with prominent yellow-black blotches at the base. Here also we found muscari, puschkinias, gageas and colchicums. On the northern slopes we found a dwarf Oncocyclus iris, possibly *I. acutiloba*, with rather small brown and purple, heavily veined flowers. A small brown and yellow fritillary may have been a form of *F. kurdica*, while we also found dwarf Juno and reticulate iris species out of flower. One of the most interesting finds, however, was quite a large white spring-flowering colchicum, close also to *C. bifolium*, with several rather large globular white flowers about three to four inches high and flowering right at the side of a large patch of snow, the lip of which even overhung some of the flowers and drifted on them. A sharp hailstorm at the same time added to the difficulties of photography.

Our best find from the area beside the lake was the slender yellow Oncocyclus iris *I. urmiensis*, sometimes described as a variety of *I. polakii* but distinct in the form of its beard. It was not in flower at the time, but flowered later at Wisley and indeed seems to have settled down there. It was already nearly ripened at the beginning of June, growing among little hard-baked red hills not far from the lake. It is the only self-coloured yellow iris that I know in this section and a very lovely plant.

Our road from Tabriz to Tehran lay over dry rocky and often semi-desert country. However, among the rocks were pale lemon-yellow holly-hocks of much beauty, dwarf blue anchusas, alliums and aethionemas as well as numerous fine horned poppies, both scarlet, yellow and apricot, the last possibly being a hybrid between the two former. At the edge of the desert country, however, we saw large expanses of mauve larkspurs, slightly smaller versions of those found in English gardens and growing with yellow nigella; also our first *Hulthemia berberifolia* which is better known to many as *Rosa persica*. It grows little more than a foot high and makes a dwarf prickly grey bush on which the deep yellow single flowers are conspicuous. At the base of each petal is a deep crimson blotch. During the summer it would become quite dry and the brushwood from it is, in fact, sometimes used as firewood. It is a handsome plant but, as suits its habitat, it is very deep rooting and consequently intolerant of any move. In this country it is borderline for tenderness and difficult to establish, and we have had little success with the seedlings which were raised at Wisley. The seed capsules, unlike those of other roses, are almost as prickly as the stems.

After a brief visit to a wonderful old mausoleum near the road at Sultaniyeh, probably the largest and also the earliest of the great blue Persian domes, and a few days spent in Tehran over necessary arrange-ments and despatch of our plants, we went up into the Elburz mountains which run like a great chain along the north of Persia between the desert and the Caspian. For our visit to the part of the range near Tehran we chose the pass over from Kharaj, west of Tehran, to Chalus where there is a good road leading up to a tunnel at the top of the pass at about 8,000 feet. An older part of the road goes right over the top as well, and this we were able to negotiate thanks to the low gears of the Land Rover; it has, however, largely fallen into disuse as a road and would probably have been impassable for an ordinary car. Colonel Goddard-Wilson, of John Brown Constructors Ltd., very kindly put at our disposal for a week a villa on the north side of the pass about a thousand feet below the tunnel, and he and his wife came up there with us and advised us for the first day. It made a most excellent base.

There is a great contrast between the two sides of the pass. The south side is covered with dry and steep stony slopes which look almost barren

from a distance. However, on these we found that there had been ixiolirions, tulips, muscari and other bulbs as well as various vetches. In one
most attractive little village to which we were led by Count Hannibal, the
curator of the native art museum in Tehran, we came suddenly on an
oasis of trees, and round his own house were hedges and banks of that
magnificent golden rose *R. foetida*, grown as it can be only in a warm
climate, five feet or more in height and half as much thick and literally
covered with its most brilliant yellow flowers each about the size of our
old half-crown. It is sad that it will not grow and flower like that in this
country. It must be quite winter-hardy, since this village was high up in
the mountains and would be under snow and frostbound for much of the
winter. The only other place where I have seen this rose flowering like
this was in central Spain where it made a five-foot deep golden hedge a
hundred yards or so along the edge of a public park. It was most exciting.

The northern slopes of the range carry more snow and at the beginning
of June there was still quite an appreciable amount, especially on the
group known as Solomon's Seat, the Takt-i-Suleiman. The moisture laden
winds from the Caspian and the clouds which rise from these have a
strong influence on the vegetation. From dry and stony scree and rock
near the top of the pass one goes down to shrubby forest, corn fields and
even rice fields by the sea and everything becomes green again. However,
the drier part near the crests of the range and the top of the pass was much
the more interesting for our collecting since it was here that the bulbous
plants grew. Near the top we found a small Juno iris just passing out of
flower and it appeared to have had olive-green and yellow flowers. This
was the Elburz form of *Iris caucasica*, rather different from the one of the
Turkish mountains; these were more olive-green in the flowers and they
had a thinner substance to the petals. It has since flowered at Wisley but,
alas, seems to have little vigour in cultivation in England. The tops of the
passes in the Elburz were naturally cold and windy areas with rather
coarse scree like the tops of mountains elsewhere, but again one felt on top
of the world. There were not many plants except a few little crucifers like
drabas and a dwarf pink tulip with pointed petals and yellow throat and
prostrate glaucous leaves. It reminded me of *T. aucheriana* but was identified as a form of *T. humilis*.

The other main tulip found on this range appears to be that described
by Sir Daniel Hall as *T. montana*. Only those with bright yellow flowers
were found in the area of the Chalus pass and they were mostly over, but
near Mt Dermavend, fifty miles to the east, a few withered petals showed
that they were scarlet and it later flowered at Wisley. The flower was
medium in size and rather globular, on stems about six inches high, and

the narrow glaucous leaves were noticeably wavy. The bulbs, which were generally deep, had a conspicuous tuft of wool at the apex.

A deep purple Oncocyclus iris with quite large flowers, although without any conspicuous darker blotch or veining, has been identified as *I. demavendica* and is probably a species new to cultivation in this country. A yellow iris, which coloured the hillsides in one area, was probably *I. flavescens* and was like a medium-sized lemon-yellow flag iris with rather pinched and heavily reticulated falls; beautiful as it was in the wild, we only took herbarium material. The large tussocks of *Astragalus spinosus* were rich in flower colour, a pinkish-purple in most cases, and seemed typical of this harsh country. A much more attractive plant was a dwarf almond, a form of *Prunus prostrata*, but certainly the best I have yet seen. The stems were not more than a foot in height and mostly hugged the rock with occasional upright sprays, and the flowers were a good rich pink and nearly the size of a florin. We only saw it once, but a small part of it is now growing at Wisley.

The predominant fritillary, possibly *F. crassifolia*, was a medium-sized one with reddish-green flowers about six inches high, heavily netted outside and chequered golden within. Though never very abundant, we found it in several areas. A scarlet *Papaver orientale*, just like the common one of our gardens, deceived us from a distance into thinking we had found an exciting and vast tulip. The ixiolirions were lovely as were also the linums, both a large white-flowered one and a sky blue, probably *Linum bungei*, which coloured whole damp meadows. An *Aethionema*, probably a form of *A. grandiflorum*, was a conspicuous plant with deeper pink flowers than we are usually accustomed to, while the same *farinosa* primula (probably *P. auriculata*), and purple orchis, probably *O. latifolia*, coloured any damp spots where a small stream might be running. The pinkish-magenta *Geranium tuberosum* was a common plant in the dry areas, while the foot-high campanula with pale blue upright flowers was not rare. A deep yellow *Arnebia* was another quite attractive rock-garden plant, while there were also dwarf yellow *Cheiranthus* and dripping yellow *Onosma*.

On the Caspian side the trees were mostly *Parrotia persica*, which must have been wonderful when they coloured in the autumn, and the chestnut-leaved oak *Quercus castaneifolia*, two plants which may be grown in English gardens, though the latter may be a little tender. It was a tremendous contrast to the southern side of the range and one changed quite abruptly. We hoped to find yellow-cushion *Dionysias* but were unlucky although on later trips Paul did find them, flowering earlier than we had been there and not so high as we had looked.

Then followed a few days of sorting specimens at Tehran and obtaining further passes, and a short, literally flying, visit southwards to Isfahan and Persepolis. These are two of the places which no visitor to Persia should miss if he can help it. Persepolis also yielded a Juno iris and a colchicum. There was still much snow, mid-june, on the great Zagros range, the grazing ground of the Bakhtiari and Quashquai nomadic tribes, running south-eastwards across the country, and this region, although drier than the Elburz, should also repay a visit. Paul and Polly Furse were able to make this visit a few years later, and other collecting parties have also been there and have brought back interesting dionysias, little cushions like aretian androsaces with almost sessile yellow or pink flowers. They seem very local to particular mountains. There were also interesting Oncocyclus irises such as the yellow *Iris meda*.

Isfahan was wonderful for the decoration of its mosques, the honey-combed arabesques of the porches, the great blue and coffee coloured domes and the slender minarets. But for me the much older Friday mosque with its much more sombre colours was the finest. Tamerlane is reputed to have stabled his horses in the courts. The great age of Isfahan corresponded to our great Elizabethan age and the buildings have been preserved intact and are very well cared for. Persepolis was quite different, one of the largest and finest archaeological sites I have seen. It lies at the base of high rocky hills from which the old city must have got its water supply. It is raised above the surrounding desert plain on a platform built with great masonry skill out of vast blocks of stone, all smoothed and cut to fit together. A modern builder would not have done better. The slender pillars rise gracefully against the orange of the setting sun, while the low reliefs of emperors with umbrellas held over them decorate the doorways of vast blocks of stone and the endless procession of tribesmen bringing tribute mount the stone staircases. At the top of the pillars were dumbbell-like cross pieces carved with griffons or other devices, and across the centre of these presumably the planks of cedarwood lay. There are no forests near now, and all the woodwork of the palace was burnt when it caught fire after a feast of Alexander the Great when he had conquered the Persians. The mighty ruins still give a feeling of the power and majesty of the Persian emperors of that time when their empire stretched from India to Greece. Roman history is discretely silent over the defeat and humiliation of the Emperor Valerian, but a relief at Naqt-i-Rustam nearby shows him kneeling and bowing to the ground before the Persian emperor, an enlarged figure on horseback with a great balloon-shaped turban on his head.

Mt Demavend, the highest mountain in Persia and only just under

19,000 feet, was our next objective. It is an old volcano with an almost perfect cone-shaped summit and carries a snow-cap throughout the year. In June the snow lay at around 12,000 feet while patches came 1,000 feet lower. It can easily be seen from Tehran when the atmosphere is clear although it lies nearly fifty miles to the east. A road takes one to a shoulder of the mountain on the south side at about 7,000 feet, and at this height there are great patches of the scarlet oriental poppy growing among the grey rocks and mingling with the deep golden *Eremurus bungei,* now renamed by some authorities as *E. stenophyllus* which grew up to about three feet. We also found it further eastwards. It was a lovely floral spectacle, especially in the early morning when the sun was low and shone through the petals of the poppy. The form was a good deep scarlet-crimson and quite without the harsh, bricky flavour of that often seen in English gardens. Another very prominent plant of these lower slopes was a large fennel. *Ferula galbaniflua,* a handsome statuesque plant up to four or five feet, with yellow stems and large heads of ochreous-yellow flowers. The foliage was being devoured by hosts of caterpillars, although the goats did not appear to touch it.

The bulbous flora was not rich and by no means as rewarding as on the main Elburz range. We found no iris nor fritillary on our climb to the snow level and only one tulip on the lower slopes; which although the petals were withered was the scarlet-crimson form of *T. montana.* This flowered at Wisley in the spring, and was a good clear scarlet with a satin sheen and a small dark basal blotch, a lovely flower. There was also a pale lemon-coloured eremurus rather smaller than *E. bungei* and later to the eastwards we also found pink and white forms.

In the higher zones the dark laval rock, almost black in places, did not prove fertile and was often quite bare but the tussocks of *Astragalus spinosus* were some of the largest and richest for colour that we saw anywhere. Even at 10,000 feet the goats had reached and were seen to crop the astragalus since there was little else in some places, keeping their heads just above the prickles and sucking hard. A dwarf golden potentilla and several bright yellow but small gageas and a deep yellow draba were common among the melting patches of snow. A pale lilac-blue veronica with very small but rather striking flowers formed small cushions in the crevices of the rock, almost like an eritrichium, but it was not such an outstanding plant. Unfortunately there was no seed.

After the visit to Demavend we began the long drive eastwards to Meshed, which is near to the Afghan frontier and also to the south-west corner of Russian Turkestan containing the cities of Bokhara and Samarkand, the mountains near which were the source of many magnificent tulips such as *T. fosteriana* and *T. greigii* and Juno irises sent to Holland by Van Tubergen's collectors in the early years of this century. We hoped that some of these plants might have crossed the border into the long range of mountains which, interrupted only by several high passes, runs as a continuation of the Elburz range up to the Afghan border. To the south of the route was the great salt desert. In patches near the edge of this grew an iris with very narrow leaves, clustered and bristly at the base. Only one was seen in flower, very palest silvery-mauve, two or three on a stem about one and a half feet in height. It appeared to be an *Apogon* iris and is likely to have been *I. songarica*, which has been described as very variable in colour.

An interesting plant of the rocks formed great hard silvery-green bosses sometimes several feet across, always on hot dry rocks and generally in full sun. There were only remains of very small pinkish flowers and we thought that we had found an interesting *Dionysia*, a genus of the Primulaceae, difficult in cultivation as the aretian androsaces and so correspondingly popular with some of the alpine gardeners. To obtain seed it was necessary to brush hard over the tussocks with a very stiff wire brush and hope that some seed might be among the debris. However, probably our labour was in vain for the plant has subsequently been identified for us as *Gypsophila aretioides* and a plant of this from our collecting is still alive at Wisley and has slowly increased in size.

We crossed the range first near Firuzkuh and went eastwards to Gorgan, the old Asterabad, and again back to the south west of Bujnurd. Unfortunately the season was by now becoming too late for the flowering of bulbous plants apart from a few alliums and pink and white *Eremurus* as well as more of the golden *E. bungei*. All were about two to three feet high. However, we were able to collect some Juno irises whose dried-up heads were still visible, and also a few tulips where a withered leaf showed. The same applied to colchicums and merenderas. Some of the latter have flowered in December and January in the alpine house at Wisley. A small pinkish-purple one, flowering without the leaves, has been identified as *Merendera hissarica* and a small white one as *Merendera trigyna*, and in subsequent years they have kept their valuable winter-flowering habit. One of the irises was *I. fosteriana* with curious mauve and yellow flowers. In spite of the dryness, tall mauve salvias, pink convolvulus and a yellow verbascum like *V. olympicum* made quite a display of colour in places. The euphorbias were also common and distinctive with masses of green or yellowish-pink flowers. A small shrubby honeysuckle with creamy-yellow flowers may have been *Lonicera iberica*.

Meshed, a place of pilgrimage for the Shiah sect of the Moslems and a meeting place for travellers from much of the surrounding districts of central Asia, was one of the most interesting cities I have ever seen, both for its peoples and its buildings. Unfortunately the mosques and the shrine of the Seventh Imam are not open to non-Moslems, but from the great ring road which the Shah Rezah built round them one could get occasional glimpses of the flashing golden dome over the shrine and its flanking golden minarets. The famous blue dome of the mosque built by Gauhar Shad, the daughter-in-law of Tamerlane, in the fifteenth century had unfortunately been taken off completely for repair.

We were allowed to camp in the compound of the former British Consulate at Meshed, a fine Georgian type house which had apparently fallen into a state of disrepair after the last British Consul had left. Sir Sacheverell Sitwell had been there years before and had written an article in one of the Sunday papers drawing attention to its sad state and I remembered this. Now, however, it had been repaired as the home and centre of the British Council representative. Both he and his wife were most kind to us and were very forbearing with me for, the first evening there, in a hurry to get as near to the mosque as I could, I had tripped over one of the little water channels along the side of the road and fallen face downwards, resulting in a black eye and a skinned and bleeding nose, and must have presented an ugly sight. However, it was no time for resting after we had come so far and next morning I needs must be out again to see as much of the place as possible. It was indeed a fascinating city, really a part of central Asia with mounted fur-capped travellers from all over the area riding about the streets. From the library of the museum, which was open to non-Muslims, one could see something of the outer courtyard. Some of this had been rebuilt and redecorated in Victorian times and looked full of 'Victoriana', the minarets having fussy decorated balconies near the top as well as the loud-speaker which helps the modern Muslim call to prayer. Fortunately the secret of painting and glazing the wonderful blue tiles had not been lost and there was still a factory for making new ones to repair where necessary, so it is difficult to tell the age of certain parts.

Our return journey to Tehran was hot but uneventful, and after a few days there, sorting and drying out our specimens in the comfortable home of Mr and Mrs F. Baxter of the Oil Consortium where we were most generously entertained, we set off for the long drive back to Ankara. This time we went southwards to Hamadan, up on to the Shah pass on the flanks of Mt Alwend and then turned northwards to Khoi and the Turkish border through Kurdistan and by Lake Rhezaiyeh on its western side. It was too late in the season and too dry to find any bulbous plants in flower now but we did collect bulbs of a Juno iris; with their old outer coats these were about the size of a good hen's egg, the largest I have ever seen for this section. These were identified as *I. aucheri* when they flowered. Tall stout stems with a massive seed capsule also heralded at least a tulip with sizeable flowers near Hamadan. The tulip has flowered at Wisley, a large scarlet flower close to *T. eichleri*.

To the west of Lake Rhezaiyeh on low hills topped with a kind of loose stony silt, almost scree, we found a fritillary, *F. kurdica*, another tulip and also a dwarf Oncocyclus iris *I. urmiensis*. Through Kurdistan the tall white hollyhocks and silvery-blue *Echinops* with large globular heads made a constant feature. Everywhere was very dry and goat-eaten until we got to the mountains of eastern Turkey where it was moister and surprisingly floriferous and full of colour – beautiful terracotta-coloured poppies in masses, mauve salvias and other labiates, almost like an alpine meadow so full of flowers was it, mostly annuals and biennials. Our way back over the Pontic passes I have already discussed. It was one of the most rewarding parts of the trip.

From Ankara I flew back with many of the plants while Paul Furze went off for a short trip in southern Turkey with Polly Furse who had flown out again. A few weeks later they drove home with most of the dried bulbs and herbarium specimens in the Land Rover. It had been a marvellous trip and I hope proved a useful reconnaissance for the later and more extensive expeditions which Paul and Polly Furse made in the years between 1960 and 1970 while I went back to work at the R.H.S. At Wisley the bulbs and plants were patiently planted into pots and next spring many flowered in the alpine house; some we had only collected with dried and withered shoots, so it was exciting to go frequently to see what had come out.

Some of the best tulips, colchicums and irises were collected in this way. These bulbs come from such different and much harder environment; under snow for months in winter, moist with soil saturated with melting snow-water in spring, then after flowering no more water and a regular hot drying-off all the summer. This does, however, mean desiccation and shrivelling of the bulb. The best method of cultivation appears to be either in pots or pans, sunk in a cold frame and kept dry all the summer from June to October, and then repotted in fresh soil in the autumn. Or better still is cultivation in the raised bulb frame where the watering can be controlled and none given all the summer. I have so far preferred to leave the bulbs for several years undisturbed, but successful Dutch nurserymen lift every year and plant in the open ground or in a greenhouse. Good drainage is of course essential and also some feeding. I like to dig in some very old and well rotted manure about six inches down when. the frames are being made and then to top dress with fertilizers such as dried blood and bone meal or even a complete artificial fertilizer each autumn. If one can arrange watering by a perforated pipe below, and never water on the surface, this seems to suit particularly well the more difficult Oncocyclus and Juno irises such as *I. persica* and *I. caucasica*, but I rather doubt if these latter will ever be really long-lived in cultivation. Still, it is a very rewarding and interesting form of gardening, and gives one flowers early in the spring and in the late winter before much is out in the open garden.

CHAPTER SIX

Sarawak and Borneo

In 1932, while still an undergraduate at Cambridge, I was lucky enough to be able to accompany the Oxford Exploration Club's expedition to Mt Dulit, a largely unexplored mountain in the interior of Sarawak. It was a great experience and probably the one which gave me a permanent taste for travel, especially to mountains. It was also a tremendously rich and beautiful area and one very fertile in new species. Out of our collectings in a few months were described forty-two new species and varieties of orchid and many in other families. The expedition was organized by Tom Harrisson, also of Cambridge, who has subsequently got to know the country probably better than any other European from his activities as a guerrilla leader during the war and later as curator of the museum at Kuching, the capital. It is a nostalgic pleasure now to jog my memory by looking back over the book which the members of the expedition produced jointly and the papers which various scientists published on the results of the expedition.

Sarawak was a unique country then, ruled by Rajah Brooke in a benevolent way, in which there was little new development and the chiefs were largely left to control their own affairs from their rather isolated long houses by the rivers, the only means of communication. They would travel along these in their magnificent long prahus, a vast kind of canoe-like long boat, while twenty or thirty boatmen would dip their paddles rhythmically to a wild chanting. As we came up river this was our first real encounter with the peoples of the interior. The outboard motor of our long boat had broken down, no rare experience, and we were waiting moored up in the dusk when gradually downstream – at first softly in the distance, then growing to a crescendo – came this chanting and a long boat manned with warriors came into view round the bend. Slightly nervous we awaited them but they could not have been more friendly and turned out to be a party sent to escort us in to the nearest long house. A visit to a long house where many families lived together was always an event and generally an excuse for a party, for the Borneo people loved a party, drinking their rice spirit or rice beer and expecting the European Tuan to drink with them. The long house was a hundred yards or so long, raised on stilts well above the ground level and entered by climbing up notched logs for ladders. The long verandah stretched the length of the house and there were separate rooms off it for each family. The long house is entirely a vegetable product – no nails, just tree trunks, rattans for rope and palm leaves for roof. It is not intended to last for centuries, since with their system of shifting cultivation the tribesmen need to move farther up river at intervals as the land gets exhausted and a new section of the forest is felled. They assume, unwisely by our modern standards, that there will always be plenty of forest; and there certainly seems plenty but it takes a very long time for primary forests, once felled, to return again to that state. However, the forests were very exciting then for a party new to them and they could be seen well from the river, sometimes even better than from inside. They made a great impression on me and I cannot now improve on what I wrote for our first joint book of essays soon after our return and so some of it is included here.

Forests, so luxuriant that they seemed a frenzy of greens, a solid wall, a never-ending skyscraper of leaves overhanging the water's edge; behind, darkness and mystery and more and more plants, many still unknown and all thrilling to the young botanist.

A new land, my first expedition to the tropics and this superb forest country – giant trees like the grains of sand on the sea-shore, their trunks hidden by the mass of foliage, the only visible stem generally the rope of a liana hanging in graceful festoons. Then, 4,000 feet above, a moss forest, where striped and gaudy pitchers hung on twisted stems and white and yellow orchids scented the air, all backed by this feather-bed of moss; everywhere, on the ground, on the tree trunks, on the branches – caves of green moss, covering tussocks and boughs, often twisted and contorted into weird shapes and bearing the semblance of strange faces.

I never felt alone in the forests. Surely it is possible for such a frenzied luxuriance to convey a thrill to the traveller, an enjoyment in such careless plenty, a feeling of unity with the life around, everywhere vigorous and healthy.

As we moved slowly up river, these forests were like a dream come true to me, a dream of abundance, of beauty and peace, but also of mystery alive behind this wall. All abstract words for concrete objects – yet the power of the forest forces is so great that we cannot be impartial to it. Man cannot fail to be dominated by the forests.

Borneo is dominated by its vegetation; all the instincts and life of the people reflect it. To understand the life of the Borneo peoples we must ourselves experience the great forest, the shapes of the leaves and the curves of the twisting lianas. The people practise shifting cultivation, and only scratch a small part of the surface of the land, burning and clearing very roughly an area in which they plant hill rice along the banks of the rivers: after a few years the padi (rice) field is deserted, and the ground rapidly becomes covered with secondary forest even denser than the primary forest which it has replaced. Probably hundreds of years must pass before the big trees grow again and primary forest is restored; no one knows for certain how long. This primary forest is the real old forest, distinguished by the abundance of very large trees and the relative thinness of the undergrowth; in secondary forest there are few large trees, and the undergrowth and tangle of lianas is much denser. It is generally possible to walk through primary forest without cutting a path; this can seldom be done in secondary growth.

One of our objects in Borneo was the investigation of the structure and species of the primary forest, of which there is little now left on the Tinjar banks. Away from the rivers, however, there is an abundance which seems unending. Looking down on the Tinjar valley from our high camp on the ridge of Mt Dulit it was possible to distinguish a sharp line of division between the primary and secondary forest – a line which ran roughly parallel to the river bank and from half a mile to a mile away from it. The secondary forest is bright green and looks not unlike a smooth lawn; indeed, this resemblance is one of the chief dangers to aviation in the tropics; I believe there are records of pilots mistaking such forests for smooth ground and attempting a landing. The primary forest is a dull dense green in colour, and the crowns of individual trees can be seen projecting above the general level even from some distance.

So dominant is the forest that it is said to be possible for an orang-outang to travel from the south to the north of Borneo without descending from the tree-tops. His only barrier would be the big rivers and, since the majority of these run north and south, they would merely prevent his spread longitudinally east and west.

Botany and plant-collecting in the tropical forests are not quite the same as botany in England. In Borneo we started with the belief (subsequently found not to be quite true) that everything floral was unknown. Still, the plants were less known than the animals. We wanted a complete collection – everything from trees to mosses. Instead of setting out with a vasculum and a knife we started with porters, large baskets and axes.

My colleague and senior botanist was Paul Richards who had already been a member of a similar expedition to British Guiana and is now a Professor of Botany in Wales. He was particularly interested in the study of tropical forests and here there were plenty. We formed quite a procession every morning, headed by Richards and myself. Then came Ngadiman, an excellent trained Malay collector, whose services had been lent us by the Director of the Singapore Botanic Gardens. He was a little man, and he always wore a most immaculate white topee. Ngadiman was followed by Lumbor, a forestry guard, lent by the Rajah; then came two or three Kenyahs or Punans carrying baskets and axes on their backs. These people always carried loads on their backs, never on their heads like the Africans. If the forest was thick, and we strayed from the main path, they would cut a path with their parangs; always when we left the main path they would cut and bend over small twigs so that we should be able to find our way back again. It is very easy indeed to become lost only a few yards from the path.

In camp we left Omar, a valuable old Malay, who dealt most skilfully with the Herbarium material. He would change the papers and keep a fire going nearly all day drying the specimens; in fact, he almost cooked some of them, with excellent results.

There is something about the rain forests of Borneo that will always lure back the traveller who has once visited them; the luxuriance, the chaos of tangled growth and the vastness of some of the trees inspire awe. A tropical forest is not dark, as is often suggested; the sunlight does penetrate through the 'canopy', and the ground level is neither so dark nor so gloomy as in a thick pine wood in Europe.

Many imagine that these deep tropical jungles form a paradise of large and brilliant blossoms. The flowers indeed are there, but they are not found in conspicuous masses as on an English heath or in an alpine meadow. If you want to see hillsides covered with colour, go to the Alps or the Himalaya, and not to the tropical rain forest, which is indeed a paradise, but a paradise of greens; there is so much of this colour and so many different shades that the effect is somewhat unreal. Often I felt that our wanderings in the forest must be only a pleasant dream and that I should wake to find myself in Cambridge, slumbering over the fire.

In the forest we had to search hard, frequently with field-glasses, to discover the flowers; it is easy to go for a long distance without seeing any at all, for they are mostly on the tops of the trees where the light reaches them; only a few fallen corollas or petals below betray their presence.

Our path was not a smooth or flat one, just a trail cut through the forest on the mountain-side. Practically all the ground between the river and the mountain consists of ridges with narrow and sometimes knife-edged tops, up and down which we perpetually scrambled, slipping and sliding on the clay soil and clinging to the frailest supports to help us up; indeed, sometimes I felt quite glad when we found a tree in flower and were able to rest while it was examined. The main crest of Dulit consists of an escarpment of Lower Miocene sandstone which runs more or less parallel to the general course of the river for some twenty miles, maintaining a ridge of fairly even height. In many places, where it was very steep, the Kenyahs had prepared for us ladders made out of notched tree trunks; sometimes these ladders were almost vertical for twenty or thirty feet, and it would have been impossible for laden porters to ascend the mountain without them. The dampness natural to the mountain and the forest soon covered these trunks with a mass of minute green plants and algae and made them extremely slippery. Since there was often no hand-rail we had to balance ourselves precariously on them. We quickly found that ordinary sand-shoes with rubber soles were much the best kind of shoes to wear, and that nailed boots slipped the worst on the ladders.

There was a small stream which wandered down from the escarpment of Dulit into the Tinjar by our camp. Our trail crossed this stream eight times. Generally it was easily fordable by stepping-stones, but after heavy rain it would become a raging torrent and form an effective barrier between the two camps. One day, for botanical purposes, we felled a large tree across one of the fords, which we thought would provide an adequate bridge. The trunk was several feet clear of the water. A few days later it was gone, and we found that it had been swept by the spate several hundred yards downstream into a quite useless position.

Along the stream we could see the size of the forest trees since it was possible to view them from a distance. Many were as much as a hundred and fifty feet in height and a hundred feet to the first branch. Near the base the trunks were often buttressed, and the buttresses would sweep out in serpentine curves from the base of the tree. The cause of the buttressing is still largely unknown to botanists, although it is probable that they do provide some extra support for the tree.

Along the stream, also, we could see best the luxuriant growth of liana and epiphyte. I think that these are two of the factors which contribute most to the beauty of the forest. The stems of the lianas would hang like bell-ropes from the tops of the trees to enter the ground far below. Lianas are a characteristic of tropical forests. They are woody climbers with rope-like stems and masses of luxuriant foliage tumbling and smothering in the tree-top level where the light is reached. They are most useful to the wandering botanist, since up to them a man can generally be persuaded to climb, exactly as up a rope, to gather specimens from the tree-tops. In England the nearest approach to these lianas is the Traveller's Joy, *Clematis vitalba*, but it displays little of the rope-like stem and the great masses of luxuriant growth found in tropical lianas.

Many of these ropes assumed weird forms. We saw one which resembled closely a corkscrew eighty to ninety feet long, winding in a spiral up to the forest top. It was probably a species of *Bauhinia*, a leguminous liana common in the tropics. Another such climber had a flattened stem twisting in graceful curves and loops, but always repeating the same series of forms and sequence of curves, even as some ornate wallpaper returns to the same pattern again and again.

The rattans or climbing palms caused as much trouble. The commonest was *Calamus*, a climbing palm with beautiful feathery fronds, but in which the leaflets at the ends of the fronds and the stems bear stout hooks; these would frequently catch in our clothes, our hair or our flesh. The stem of this palm often attains a length of several hundred feet, and when stripped is largely used by the people as rattan for fastening round joints in their houses and making baskets, etc. Although slender, it is very strong. It is the rope of Borneo as banana fibre is the rope of East Africa.

The high temperature and the high humidity of the tropical forests encourage growth of all kinds; but to obtain light it is necessary for a plant to grow up quickly to the top of the forest as a young tree does, or else to grow on some support such as the branch or the trunk of the tree. These are the epiphytes – orchids, aroids, ferns, mosses and many other groups. They are not parasitic on the trees, but are merely supported by them. Round their roots they collect leaf-mould, and from the damp air they draw in moisture. This is especially true of many of the epiphytic orchids whose roots are surrounded by a layer of spongy tissue called the velamen, which acts like a sponge.

Epiphytic on the leaves on many plants, including other epiphytes, there would often be small masses of liverworts; merely a green form to the naked eye, but when looked at under a lens a delicate and beautiful structure of slender branches and green plant body. The epiphyllous liverworts seemed to me to represent the acme of epiphytism, the last word in sponging on your neighbour.

We had been asked especially to look out for and to collect as many of the flowers of the trees as possible. Often the flowers of the trees are the last flowers to be collected in the tropical forest, since they are not often visible from below and are difficult to obtain. Whenever a few flowers were found on the ground, we would settle down to a 'tree conference'. This became an established part of our routine and would last from five to twenty minutes. It provided for me an unfailing source of amusement. It was no easy matter to decide from which tree the flowers came. Everyone would peer upwards into the 'canopy'; field-glasses would be passed from hand to hand until the canopy had been inspected from every angle except the one vital one – namely, from above. Ngadiman and Lumbor would go round, blazing all the trees, examining the latex, smelling the bark and sap. At last one of them would come to a decision; nearly always they were correct. We ourselves could seldom see any flowers above.

If it was possible for a man to climb the tree or by a neighbouring liana to reach the flowers, this was done. We had one man who was a champion climber. I am sure that he would have put many of Britain's best gymnasts to shame. But if flowers could be obtained in no other way, we would cut down the tree. Axes would be pieced together for the head was generally carried separate from the shaft; the men would start work and down the tree would come, often in an amazingly short time as the axes were very small and the wood of many of the trees very hard. The wood of bilian, one of Borneo's finest timbers, is so heavy and close-grained that it will not float in water. Always the fall of a big tree fascinates me. In these Bornean forests it was magnificent, since the tree-tops were so thick that one tree could not fall without disturbing others; for several moments after the main crash, little pieces would continue to fall, while the sound echoed round and round from tree trunk to tree trunk; then, by contrast, came a deeper silence than that to which we were accustomed, until gradually the smaller noises of the forest began to assert themselves again. One forest giant would generally bring down several smaller trees and lianas in its wake, and we felt disappointed if we could not collect considerably more than the flowers of the actual tree for specimens.

Systematic science demands many sacrifices in the way of fine trees, flowers, birds and animals, beautiful butterflies and strange insects, but in the case of the tree, the fall of one giant provides light and space for seedlings to grow up and his place is filled again.

While Richards measured the trunk of the tree for diameter and height, and took out a longitudinal section for a timber specimen, I would scramble about among the debris and collect epiphytes, particularly orchids, of which we obtained some interesting species in this way. The epiphytic flora of a great forest tree is enormous. I have often thought that it would repay an ecological study. Along one branch alone there may be hundreds of plants, and with them are collected humus and mosses and insects dependent on the plants. A single tree-top is like a small world of its own.

There are, however, a few trees in the tropical forest which do not produce their flowers on the top branches, but out of the old wood of their trunks often only a few feet from the ground. Such trees are peculiar to the tropical forests. I don't think that any occur in England or in temperate countries.

SOME ALPINE FLOWERS

(Top left) *Pulsatilla alpina* absp. *apiifolia.* The yellow Alpine Anemone

(Top right) *Gentiana verna*

(Centre) *Campanula linifolia* with the Jungfrau in background

(Bottom) *Soldanella pusilla* in the Dolomites

SOME MEDITERRANEAN FLOWERS

(Top) *Spartium junceum* in Corsica

(Centre) *Euphorbia wulfenii* in S. Greece

(Bottom left) *Anemone pavonina* near Sunium in Greece

(Bottom right) *Paeonia clusii* in the White Mountains in Crete

SOME FLOWERS OF
THE PACIFIC STATES
OF N. AMERICA

(Top left) *Lilium
washingtonianum* var.
purpurascens

(Top right) *Erythronium
grandiflorum* on Mt.
Rainier

(Centre) *Penstemon rupicola*

(Bottom) *Lewisia tweedyi*

SOME FLOWERS OF
THE NEPAL
HIMALAYA
(Top) *Primula sharmae*

(Centre) *Arisaema
wallichianum*

(Bottom left) *Rhododendron
cowanianum*

(Bottom right) *Iris
kamaonensis*

The almost invariable association of the larger epiphytes with ants caused me much discomfort, and it seemed that the smaller the ant the more painful was the bite. Luckily the effect only lasted for a short time, but frequently I had to suspend my collecting while we brushed and picked the ants off. There are several epiphytes which produce specialized internal cavities in which the ants live, such as *Myrmecodia*, a plant in which the base is formed into a large tuber, honeycombed with passages inhabited by ants. There is also a *Polypodium* fern which has a great flattened base hollowed into chambers. It has been suggested that the ants protect the plant from other insects, but it seems more likely that their main importance to the plant lies in the humus which they collect. There are also species of *Macaranga*, plants belonging to the spurge family, which contain hollow cavities for the ants in the flower-stalks. There is a small hole by which they can enter and leave the cavity. The exact relation between the plant and the ant is not known.

We were particularly keen to find orchids which might prove of some horticultural value, and indeed several that we found were of considerable interest for their bizarre form and delicate colouring. I found them more interesting by far and sometimes more beautiful than the gigantic and showy hybrids produced in such numbers today. For pure show, however, few of the Bornean species could compare with the products of modern orchid hybridizers. The orchid flora is very rich, and undoubtedly many species yet remain to be recorded. Richards, who had also visited British Guiana, reported that the flora of that region seemed even richer in orchids of horticultural value than the flora of Borneo; this would easily be confirmed from an inspection of any orchid exhibit at a flower show, since all the *Odontoglossums* and the *Cattleyas* are American in origin.

The prevalence of epiphytes, particularly orchids, seems to be closely related to the roughness of the bark. In the case of a very smooth bark little humus or moisture collects on the branch and few epiphytes are found, while in the case of trees with rougher bark the higher branches are often covered entirely with epiphytes which take advantage of the humus and moisture collected on them.

Scent is a frequent characteristic of the Bornean orchids. There was one small *Coelogyne* with white and yellow flowers common in the moss forest. It had a very sweet smell. I could always tell when a plant was near by this means, and would then search until I found it. There is another Bornean orchid, *Bulbophyllum beccari*, which is reputed to have a smell far stronger and far more unpleasant than that of any durian, which can be smelt fifty yards off. The flowers are reddish-brown and are, I imagine, fly pollinated. Nearly all plants which are pollinated by flies smell like bad meat; we have only to think of *Aristolochia* or *Stapelia* to confirm this. The leaves of *B. beccari* are magnificent. It is probable that they are the largest of any orchid leaves, being entire and over eighteen inches in length by twelve inches in width. The stem is almost wooden and ascends the tree trunk, clinging with small roots like ivy, while the vast leaves grow out as brackets, collecting a mass of leaf-mould at their base.

In extreme contrast to this plant are the minute flattened stems of *Taeniophyllum*, which we frequently found on the highest branches of the trees. Here there are no leaves, and the stems resemble slender green tapes creeping over the branches. The flowers also are minute.

On a tree trunk a little distance from the trail, I found one of the finest orchids we saw in Borneo. The flowers were large, about two inches across, and borne singly, one on a stem. They were pale buff-brown in colour, with orange flecks and lines. The lip of the orchid was poised and curved forward, so that it moved in the wind; the lateral petals were like two superb orange-brown moustaches, emerging above a high stiff collar. There was a large mass of this plant. We took half, and I still remember Ngadiman walking down the mountain with it, supported like a wreath round his neck. Subsequently this orchid flowered in England in Sir Jeremiah Colman's famous collection and received an Award of Merit at the Chelsea Flower Show in 1938 under the name of *Bulbophyllum lobbii* 'Gatton Park variety'. It was unfortunate that the name suggested a hybrid raised at Gatton Park rather than a species collected on Mt Dulit in Borneo. I do think that such varietal names should not be given to species collected in the wild, but that all names for such plants should, as far as possible, be either descriptive of the plant or be derived from the place whence it came.

As well as these fine orchids we found a number of species of *Cirrhopetalum*, with deep crimson, reddish-brown and orange small-hooded flowers, from the base of which extended a long strap-shaped labellum. These flowers were grouped together at the end of long stem, so that they often formed a semicircle of colour a couple of inches across, delicately poised and swaying in the wind.

Grammatophyllum, the largest orchid in the world, grows in Borneo. We found a clump on the edge of a tree by the rice-field clearing. At first I thought it was a kind of palm, so vast were its long filamentous leaves, borne on a thick stem often six or eight feet in length. Unfortunately we did not see it in flower here, but I later met it in the Botanic Gardens at Buitenzorg (Bogor) in Java, perched in a tree of the great *Canarium* avenue. The flowers were orange and brown, several inches across, and borne on great six-foot spikes, while the leaves must have been an equal length. Occasionally plants are seen in England, but I have heard that it very seldom flowers here. Another of the astounding orchids of Borneo is *Arachnanthe (Vanda) lowii*, whose flowers are borne on long garlands often thirty feet in length. They are crimson and brown and about two inches in diameter, but the two basal flowers of each spike are absolutely different from the others, both in colour and shape. They are bright yellow and have small crimson spots. This wonderful plant is named after Sir Hugh Low, an early resident magistrate who made large collections.

All these orchids of the forest require to be grown in a stove-house in England, but the orchids of the moss forest into which we suddenly entered at 4,000 feet may well be grown cooler.

Although many of the orchids are epiphytic on branches or trunks of other trees, they are not parasitic. We found many interesting parasites, however. Once our collector found buds of the giant parasite *Rafflesia*, the largest flower in the world, but unfortunately, together with them, he cut the root of the liana on which they were growing. Severed from their host, they did not open, and we never found any more.

Perhaps the most brilliant floral sight in the forest was presented by the flowers of a member of the *Loranthaceae*, a parasite allied to the English mistletoe. They had a brilliant red and orange tubular perianth and covered the whole crown of a great tree with colour.

Some of the most interesting trees of the mixed forest were the 'strangling' figs, the seed of which is carried by birds and germinates in a fork among the top branches of some tall tree where the fig grows epiphytically. Long roots are produced which often clasp the trunk of the 'host' tree and kill it in much the same way as ivy kills trees in England. They reach the ground and enter the soil as normal roots; thus, by the time the 'host' tree is dead, the fig is independent. A bizarre and fantastic appearance is frequently presented by these roots, which are twisted and gnarled, while a forest of aerial roots (which grow straight for one hundred feet from tree-top to ground) is produced when the tree gets older.

Perhaps, more than any other plant, palms are associated with the tropics, and Borneo was full of beautiful palms. As we came up river in the brackish estuaries, the banks were lined with the Nipah palms, wonderful great fronds, thirty feet of glorious feather, springing directly out of the mud, waving in the breeze, glowing orange in the evening light. Then in the forest there were the Rattans, aggressive climbing palms; occasionally we saw one of the beautiful sago palms, dignified trees with a twenty-foot trunk surmounted by great waving plumes. The sago of the pudding comes from the pith. In the actual undergrowth of the forest we would find the Licuala palms, plants with no central trunk and large fan-shaped leaves. These were used often for roofing. Our high camp was roofed with them and the sun could be seen through it in places. It did leak in heavy rain, although not too badly.

Ferns were also plentiful in the forest, ranging from the great clumps of *Angiopteris*, akin only to the primitive monster, to the Filmy Fern, so ethereal that its fronds were transparent, being often, I believe, only one cell thick. The Bird's Nest Fern generally perched precariously, attached to the side of tree trunks, the fronds surrounding a perfect nest.

Among the fungi we found many closely resembling English species. One of my most exciting finds, however, was a species of *Dictyophora*, fantastic and exotic in its appearance. The base resembled a delicate hair-net, pale pink in colour, suspended from a central fleshy stalk which ended in a porous yellow 'head' apparently secreting some nectar, since even large insects were continually attracted to it. Ten inches in height, this fungus was a magnificent sight, almost a surrealist object, growing out of an old prostrate tree trunk.

Above 3,600 feet we suddenly entered a different world. It was the moss forest, a weird, a fantastic zone, where everything was covered deep in green cushions and tussocks of moss sponges — moist, squashy, frightening but beautiful. The transition from the 'mixed' jungle type of forest to the 'moss' forest is quite abrupt and presents one of the problems which the results of the expedition have not so far explained. At the bottom of the ladder up a small cliff was ordinary forest, at the top was moss forest. Drifting cloud is probably a factor in the formation of the 'moss' forest, but no certain correlation can be made of the sudden change between the two types of forest and the cloud level. It also seemed to us to become suddenly cold as we entered the moss forest. Dr Hose, a famous Sarawak resident officer, indeed compared the weather in the Dulit moss forest to a bleak November morning in England. The effect was accentuated by the contrast with the heat below, and we were all glad to wear sweaters and often tweed coats as well.

The 'moss' forest consists of small trees, very few of which are more than forty feet high: they present a fantastic appearance, being gnarled and bent and covered in the lower parts with a dense growth of mosses and liverworts, hanging in thick mats and long festoons from the trunks and the undergrowth; although the forest is called 'moss', liverworts actually predominate. This growth is generally a foot in depth, and in some places is as much as ten feet, giving the appearance of green grottoes or fairy caves, forming arches and wreathes from branch to branch, while green pillars and stalactites join floor to roof. Pitcher plants, orchids and rhododendrons grow here in profusion. Many were new to science, and they give more colour to the moss forest than there is in the forests of lower levels, where the chief source of colour was the mauve and purples of the young leaves, drooping delicately and bashfully from their stems.

There is a curious contrast between the luxuriance of the moss with its general dampness and the leaves of the trees, the majority of which are small and often ericoid (heather-like) or thick and leathery in form. It would appear difficult to correlate such apparently sclerophyllous features (features of plants growing in very dry or very acid environment) with the prevailing humidity, although it seems that the reason must rest in factors of soil and light and acidity. But this last factor is more a result of the moss than the cause of it. In general, more light penetrates into the moss forest than into the mixed jungle forest. There are great contrasts in this respect which add to the mystery. The sclerophyllous and ericoid type of leaf is particularly conspicuous on the tops of the peaks where the exposure to sunlight is greater and the growth is dwarfed, while the plants are slightly different from those of the general moss forest

In the early morning the dewdrops sparkle on the feathery mosses, and the filmy ferns are almost ethereal in their delicacy; then it does not require much imagination to conjure up little winged figures peeping out of the grottoes made by the moss. There were mosses growing here on the ground which might have been miniature Christmas trees, hung with fairy bangles when the dew flashed with light on them. The pitchers would have provided drinking-cups, nestling in the moss or swaying in the wind.

It was indeed an enchanting place, although there always seemed something mysterious and uncanny lurking among the fantastic shapes of the trees and moss, and it does not seem strange that many of the Borneo peoples regard the mountain-tops as the home of the spirits of the dead.

In the 'moss' forest flora, Australian types, such as *Casuarina, Dacrydium* and *Leptospermum* predominate, while in the lower forest, Malayan types are easily dominant. It is pleasant to speculate that the 'island' mountain summits of Borneo are relics from the day when there was a land-bridge between Asia and Australia – islands which survived the inundation of the lower lands which now surround them. I fear, however, that such a statement can be little more than a speculation. There is Wallace's line of deep water just east of the island, while there is no such line separating Borneo from Malaya. This moss forest differs completely from the lower rain forest both in structure and in the actual plants; indeed, out of the several thousand plants which we collected, so far only fourteen species have been found common to both the lower rain forest and the moss forest.

Nestling against the tussocks of moss were the pitchers of *Nepenthes* – Nepenthe, the old goddess of sleep and oblivion – and certainly it is oblivion for the many insects which find their way into the pitchers and are drowned there and slowly digested, for the *Nepenthes* are insectivorous plants. But they are more – they are beautiful plants; their pitchers are streaked and painted with a theatrical brilliance, their form designed by a Cellini endowed with a Machiavellian and wholly diabolical cunning.

As we emerged into the moss forest we came upon a pool into which a little waterfall continuously poured, throwing out endless ripples to the green edges. Near here we found our first moss forest pitcher – *Nepenthes tentaculata*. It was one of the smallest, but one of the most beautiful. The pitchers are borne at the ends of the leaves, dangling on a curved stem, which is a prolongation of the mid-rib of the leaf. It is hard to describe their form; they are like some very elaborate and exotic pipe, coloured on the outside green and streaked with crimson. They are variable in colour. Often they have a bluish-purple tinge. The inside of the pitcher is pale blue. Often six inches in length and two inches in diameter, they hold a considerable amount of liquid. The angles and the lid are feathered with deep crimson hues. Always these plants grew in shade and seemed to like a lower light intensity than the other species we found, which clambered up to the light.

In Borneo there is no doubt about their insectivorous habits, but it is doubtful whether the insect food is so necessary to them. In English greenhouses they seem to grow quite well without any insect food. Perhaps we may regard the decayed insect food as a savoury titbit, supplying extra nitrates and other salts, to use an anthropomorphic simile. In nearly all species the rim of the pitcher is smooth and extremely slippery so that any insect, attracted by the bright colour or by the honey secreted by the glands around the rim, would be inclined to fall down into the fluid below. Some of the larger pitchers contain more than a pint of fluid. One-way traffic only is ensured, as the rim of the pitcher is generally formed after the manner of one of those unspillable ink-pots and has stiff hairs projecting downwards from the edge. There is also a lid which does not close but against which any insect which attempted to fly out would be likely to collide.

The inside of the pitcher is covered with small glands which secrete a fluid, allied to the proteolytic enzymes of our own insides and possessing digestive properties. The fluid inside the pitcher is distinctly acid, and it is probable that the digestive functions of the enzyme can only work in an acid medium. Even a young, unopened pitcher contains some acid fluid. In a large pitcher the greater part of the fluid is probably water which has condensed inside the pitcher or entered as rain. Inside the pitcher are remains of many kinds of small insects in varying stages of decomposition, mostly small flies, beetles and moths, and occasionally parts of large insects. But, an amazing fact, the pitchers also contain a considerable fauna of living aquatic insect larvae, particularly mosquito larvae. It seems probable that the digestive powers are weak, as the acid is considerably diluted with rainwater. It has also been suggested, and there is some experimental evidence to support the suggestion, that these mosquito and fly larvae contain an anti-protease substance which would inhibit their digestion by the proteolytic enzyme of the pitchers.

As we cut our way through the moss forest we found other remarkable species of pitcher plants, *Nepanthes rheinwardtiana*, *Nepenthes veitchii* and *Nepenthes stenophylla*. I think that *N. rheinwardtiana* was the most graceful and beautiful of all that we met. The slender stems scramble up through the moss and small trees to reach the light. Often they are thirty feet long, and at the top only are found the large crimson pitchers dangling free in the air, ten or twelve to a plant. As the leaves die off, so do the pitchers, and more grow above on the young leaves. In shape they resemble narrow flasks and are often as much as fourteen inches in length. They arc not streaked and blotchy as other species, but a uniform rich deep crimson in colour. Inside, the pitcher is a pale green in colour, while just below the cup are two brilliant emerald spots which gleam as eyes.

The pitchers of *Nepenthes veitchii* are large, and resemble both in shape and colour the popular hybrid often seen in cultivation, and named after Sir W. Thistleton Dyer. It is a magnificent plant, a flamboyant beauty. The pitchers are covered thickly with a down of pale pink hairs, while the lip of the mouth is prolonged upwards into a fan-like structure of extreme slipperiness, coloured with brilliant diagonal stripes of green and scarlet. They are borne on rigid stems which adpress them closely to the tree trunk. The stiff leaves also clasp closely round the trunk.

Nepenthes stenophylla was distinctive of the slightly more exposed and drier situations on the very top of the various summits of the Dulit range. The pitchers are large, graceful in shape and brightly coloured, streaked with raw-meatlike crimson and scarlet. At the apex of the pitcher where it joins, the mid-rib is always a little spiral twist, somehow comic and delightful.

In the moss forest we also found many beautiful orchids. The most common was a *Coelogyne* which proved to be a new species. It had white and yellow scented flowers in a long pendant raceme which hung down on the moss like a necklace. It was so common in some parts that we could often smell the flowers before we actually rounded the corner and found the plants. We also found a fine *Dendrobium* with large white and yellow flowers and a most curious terrestrial orchid, *Corybas johannes-winkleri*, with a small deep crimson and white cup-shaped flower from which protruded a crimson lip shaped like a ledge, while from the rim of the cup extended three long filaments like whiskers. This grew out of the base of a single heart-shaped and beautifully-veined leaf.

My first day in the moss forest I got a surprise. Looking up at the fork of a big tree I saw a brilliant orange flower. I sent Arwang, one of our Kenyah helpers, to fetch it, and he came down with one of the most brilliant epiphytic rhododendrons that I have ever seen. Later, we found another kind with fine shell-pink flowers, and yet another with scarlet flowers. They all grew as epiphytes and were rhododendrons of the type we sometimes see in hot-houses labelled Javanese hybrids. But they were as fine as any of their type I have seen in cultivation. Some of these were forms of *R. brookeanum*, which is also found on Mt Kinabalu in North Borneo, while others may have been forms of *R. javanicum* itself. They were lovely clear colours and the flowers were thick and waxy in a loose truss. Others included the rarer *R. crassifolium* with apricot-orange flowers, each lobe about three centimetres long, and *R. durionifolium* with yellow or orange flowers of about the same size. I have not seen either of these in cultivation. We collected some seeds but unfortunately they did not germinate in England. On the mountains of New Guinea there are also fine rhododendrons, perhaps even finer than these, and from there it is conceivable that some might be hardy in English gardens. Some might even be tolerant of lime, since many of the New Guinea mountains are made of limestone with innumerable knife-edged ridges.

From the moss forest we passed over the ridge of Dulit down into the forest by the stream of the Koyan. This was quite different in appearance to that we had found in the Tinjar, and we found many new plants.

The most unusual feature of this forest is the soil, which is almost pure sand, and undoubtedly this is the limiting factor in controlling and forming this type of forest. A curious feature was the absence of buttressed trees, although with such a light soil there would appear to be a greater need for them than in the 'mixed' forests where the soil is a heavy clay or loam. The ground level here gets more light, and there is a much more dense undergrowth. There was a definite tendency towards species dominance, and *Agathis alba*, the dammar, produced thirty-five per cent. of the trees over sixteen inches in diameter. This tree, a broad-leaved conifer, sometimes reaches a considerable size; it produces a valuable resin which is collected by the people and exported for use as a colourless varnish.

Richards told me that he had found a somewhat similar type of sandy soil in British Guiana. The species and genera of the plants were different, but the general appearance was similar. In the same way there was a certain similarity in appearance, although not in species, between the moss forest which I saw here in Borneo at 4,000 feet on Mt Dulit, and that which I saw later at 10,000 feet on Ruwenzori – the Mountains of the Moon – in Central Africa. There, also, was the abrupt change of vegetation as we passed from one zone to another on the mountain.

In this Koyan sand forest we found more pitcher plants, and they were in many cases the same species as we saw in the sand forest by Marudi on the great Baram river, and in the scrub sandy country inland from Miri, which is the oil centre near the mouth of this river.

Nepenthes bicalcarata had perhaps the largest and the most magnificent of all the pitchers we met; in size it is only rivalled by *N. rajah* and *N. lowii*, from Mt Kinabalu, the largest in North Borneo. The pitchers are nearly globular and often were as much as six inches in diameter across the mouth. They vary in colour from a pale green to a deep crimson. It is a vigorous species, and several plants were found fifteen feet in height. The most exciting part of the pitcher is two stout spines which project downward from the lid and are very prominent. There is a story that one eminent botanist, desirous of hoodwinking the public, affixed a dead rat to these spines and proclaimed the plant as a mammal-catcher. Normally, however, insects form its only prey, while it is fairly certain that the plant can grow quite successfully without any insect food, living as other plants do. Here also were the red-streaked *N. rafflesiana* and the paler green *N. leptochila*. In the swamp part of the Marudi forest we also saw *N. ampullaria*, a species which has clusters of small pitchers in groups one above another round a stout stem ascending to the forest top. These pitchers had no lid, but a small strap-like handle sticking out from the side. They would have made splendid cups. These plants were growing in a swamp forest where the trees formed acrid roots sticking up from the ground and covered with warty encrustations.

Seeds of some of the species of pitcher plants collected in Borneo have been germinated successfully in England, but their growth is very slow. The atmosphere must be uniformly damp, but different species seem to like different amounts of light. From the extreme localization of their distribution in the field it would seem that *Nepenthes* are very closely related to their environment and that any successful grower must follow these conditions as far as possible. I doubt also whether the species from the sand forest habitat and those from the moss forest should even be grown together in the same house. At the Singapore Botanic Gardens I was told that they could not grow mountain species of *Nepenthes* successfully, but had to send them to the garden on Penang Hill.

In describing tropical rain forest, the terms 'stratification' and 'canopy' are frequently used, not only by botanists, but also by novelists and the more sensational travel writers; in their application to the 'mixed' forest of Borneo it seems that they are somewhat misplaced, and indeed incorrect. Stratification suggests a definite break between the layers, a discontinuity, while in the 'mixed' forests there was no definite break at any level nor was there any sudden change in the gradient of temperature or humidity from the tree-tops to the ground. The undergrowth and the trees tend to arrange themselves in somewhat vague layers, but there is a gradual and not a sudden change from one layer to the other, since much of the lower layers is composed of young plants of the species which, when mature, occupy the higher layers. The only layer which could possibly be called a 'stratum' is represented by the crowns of the tallest trees about a hundred and ten feet in height, the base of which was often separated from the tops of the crowns in the lower layer by a gap of a few feet. These crowns were not in contact laterally and so did not present a continuous 'stratum' or a 'canopy', which suggests a closed and a flat covering. Nevertheless, it was easy for an orang-outang to jump from one to the other. Numerous flecks of sunlight penetrated through to the forest floor, while the undergrowth was very uneven, doubtless due to this reason. The crowns of the second layer of trees, which average about sixty feet in height, are frequently in contact with their neighbours and intertangled with lianas, and this layer has the best claim to be called a canopy, but it is neither completely closed nor is it confined to any definite level, but grades imperceptibly into the layer of small trees below, while the undergrowth layer is continuous with this. This idea of structure was gained by clear felling and measuring the trees on sample strips of forest. The ornithologists also report that there was no sort of distinct stratification of bird life in the forest.

The botanical procession normally started out about eight o'clock and walked and worked till lunch-time. We would eat lunch seated on some old log or on the ground. Lumbor would often manage to cook a hot meal, lighting a fire with amazing dexterity, although all around the wood appeared to be damp.

One of the men would prepare for us a small cup constructed out of palm leaf and made into the shape of a box. As a drinking-vessel it proved excellent. We would return to camp generally about four, and then would come the serious work of laying out our specimens and writing up notes on them. On to each a number had to be tied corresponding to the number in the notebook, in which were entered details about habitat, growth, abundance, colour of flowers, etc.

In the stream by our camp there was a most charming bathing-pool where we would often swim of an afternoon or evening, apparently quite secure from crocodiles. I also used the stream, but a little lower down, for developing photographs, finding the water cooler and cleaner than the water of the main river. Photography was an important part of our job. There I spent many evenings before dusk sitting on a rock by the side of the torrent and meditating, while at intervals shaking the developing tank. Returning alone to the camp I would generally be rewarded with the sight of more animal life than when I moved with a party. Great hornbills could sometimes be seen flapping about in the tree-tops, while a small squirrel would peer out of the undergrowth at me, or a vast squirrel, with a brush like a fox, would hurry across the path. I would also be cheered by a series of fairy lights, dancing at different levels in the forest. These were the entomologists' light traps, which were suspended from several large trees in the neighbourhood of the camp. They hung in chains, each trap at a different height, and it was hoped in this way to gain some idea of the zonation of insects from the ground to the tree-tops.

Many gifts of plants and insects were brought to us by the Borneo peoples, and the majority of these, particularly the insects, were striking and bizarre; it will be a long time before I forget Hobby's noble exclamation of 'Oh, joy!' when a moribund and damaged specimen of the commonest insect was brought to him in the middle of lunch and required immediate attention. The man was invariably gratified by such acclamation of his gift and would then bring another next day, generally at the same time. I fear the botanists could not rise to such hilarious gratitude; yet we tried to show our appreciation, frequently with the gift of small quantities of coarse tobacco. A few good plants were obtained in this way, but they were generally accompanied by insufficient data as to habit of growth and location.

Once, expecting a flower, I was presented with a small water-snake in a paper bag. The situation was, however, my fault since I had failed to understand the name given to the contents of the bag, which had been introduced as 'ular bungah'. 'Bungah' I knew to mean 'flower' and so expressed a desire for the gift, unfortunately failing to comprehend the 'ular' or snake part. However, when the bag was opened, I managed to move hurriedly to another part of the boat, and the snake jumped overboard, where it swam about quite happily amid general laughter. I was not really sorry to see it go, since I had no material for preserving snakes on that trip.

Through all my forest wanderings flitted the most beautiful butterflies that I have ever seen, and *Ornithoptera brookeana* was the finest of them all. Moving through the tree-tops and over the edges of the cliff they seemed more like small birds than butterflies, as indeed the Latin name implies. Their wings are covered with peacock blue-green scales, which shimmer and gleam in the sun. This insect seemed like a flower that had taken to the air, a symbol of the many beautiful things of Borneo the native bird-winged butterfly of the white Rajah Brooke.

CHAPTER SEVEN

Mountains of the Moon: East African Equatorial Mountains

The most exciting and peculiar plants that I have ever seen or indeed ever expect to see, are the giant *Lobelias* and *Senecios* of the Equatorial mountains of East Africa. They are plants of such extreme personality, and still so little known in England, that I make no apology for devoting a chapter to them in this book, although I have already written about them at greater length in my former book *Mountains of the Moon*.

It is surprising that not a single one of the plants from these mountains is in general cultivation in England today, either as a hardy plant or as a greenhouse plant. Coming from altitudes very similar to those at which many of our most successful garden plants have been collected in the Himalayas, we might have expected a different story.

The Equatorial mountains of East Africa include Ruwenzori 16,794 feet, Mt Elgon 14,172 feet, Mt Kenya 17,036 feet, the Birunga or Mfumbiro volcanoes (Muhavura 13,483 feet), the Aberdare mountains (Kinangop 12,245 feet), and Mt Kilimanjaro 19,313 feet.

During a year spent in East Africa before the last war when the countries were still governed paternistically as British colonies, I was fortunate enough to be able to take part in an expedition sponsored by the British Museum (Natural History) together with Sir George Taylor, lately Director of Kew, and Dr F. W. Edwards, a great entomologist from the Natural History Museum. We visited several of the mountain groups, but I was able to stay longer and visit other mountains, including one range Mt Elgon, at two separate seasons which I did with an old friend, George Hancock, then starting a new job as biological tutor at the East African University at Makerere College. Stuart Somerville the artist and John Ford, who had been with me in Borneo, also came with us. It was an outstanding experience.

All these mountains stand up like islands from the surrounding warmer plain, islands of peculiar vegetation which would seem to be a relic of a flora formerly much more widely spread. It is a curious fact that many of the genera on these mountains, and even a few of the actual species, are

similar to those found in England. High up on Ruwenzori we came upon
a white sanicle similar to that seen in many English woods while on Elgon
there was a little violet very like our wild English violet, though without
any scent. The same curious affinities were observed in the insects,
particularly the insects associated with the giant plants. Both plants and
insects are totally different from those found in all other parts of Africa,
with the exception of some of the higher parts of Ethiopia. We saw
groundsels (*Senecios*), swollen and distorted with woody trunks twenty
feet in height, lobelias like gigantic blue and green obelisks, heathers
mighty as great trees. Most alpine plants are reduced to extreme dwarf-
ness, but these have rushed to the opposite extreme and exhibit an
exaggerated gigantism.

A grey mist made a fitting background for the most monstrous and
unearthly landscape that I have ever seen. Vague outlines of peaks and
precipices towered around us. Here were plants which seemed more like
ghosts of past ages than ordinary trees and herbs. They appeared as a
weird and terrible dream to me, a botanist and hunter of strange plants.
It all seemed unreal, like some imaginary reconstruction of life in a long
past geological age, or even upon another planet. Our own familiar
common herbs seemed to have gone mad. Although not lunar in fact, they
well lived up to that name in appearance. On the ground grew a thick
carpet of mosses, some very brilliant yellow, others deep crimson in
colour. Every shade of green was represented and the tree trunks were
also clothed in thick moss, often tussocked into the semblance of faces. It is
good to be able to escape sometimes from the ordinary world; this strange
mountain carried us into a dreamland which was often a fairyland,
occasionally a nightmare. We were standing almost on the Equator, yet it
was as cold as a really cold winter's day in England, and a little ahead
there was permanent snow and ice.

It seems probable that in former geological ages there were periods in
which central Africa was very much colder and wetter than it is at present,
and that the lakes extended over far greater areas. In such a climate
types of plants and animals which now inhabit the temperate, and even
the sub-arctic regions, could live comfortably right on the Equator in a
land which is now much too warm and dry for them. Then the glaciation
in the north retreated. The icecaps on the African mountains also de-
creased, and in some cases, for example Mt Elgon, they actually disappeared, maybe at the same time as the Ice Age in Europe retreated, maybe
later. As the icecaps retreated up the mountains we may assume that the
cold-loving plants followed them until they reached their present positions. The mountain floras do not today really begin till a height of 7,000

feet is reached. Thus we may assume that the different mountain floras became isolated from one another and evolved separately. Each mountain top is a little garden where evolution may have proceeded uninfluenced by the rest of lower Africa. But allowing that all this is true, and it is difficult to think of any other explanation of these curious floras, we would expect to find fossils allied to present-day mountain plants in the lower lands. As far as I know none has so far been found, but this cannot be taken as proof that the plants were never there.

The mountains are separated from each other by a considerable distance, and there is no present bridge by which plants could pass from one to the other. Yet on each mountain there is the same general aspect and zonation of the flora, the same peculiar gigantism in certain genera, although there is variation among the actual species. This present variation would make distribution of the plants by migratory birds unlikely as a factor that acts now. Only one species of giant lobelia is common to all the mountains. This was the lowest and also the largest species that we found – *Lobelia gibberoa*, and I still grow small plants of it in my greenhouse, plunging them outside in the summer. Unfortunately it is the least decorative species, but in a full-grown specimen the spikes grow like cathedral spires up to twenty-nine feet and the leaves, pink-midribbed, may each be two or even three feet long. Unfortunately it is not hardy in English gardens and will not stand any frost since it grows in Africa below the frost line and not above 10,000 feet, where the alpine zones begin. The flowers are greenish-white and partially hidden by green linear and pendulous bracts which are larger than the flowers. It has flowered in the temperate house at Kew after about six years' growth. The nearest comparison to it in form lies in the giant blue and pink echiums from the Canary Islands which grow so well and seed themselves in Riviera gardens – or possibly the giant *Eremuri* of Persia and Turkestan. I think it is significant that the great variation occurs higher up the mountains. Thus these particular plants open up innumerable questions which are still unsolved. How did they get on to these mountains, or did they develop there? What caused their extraordinary gigantism? How fast do they grow, how long do they live, and under what conditions do they now subsist?

On the origin of the gigantism no one had been able to produce any really satisfactory theory. We can only say that it would appear to be due to the complement of the rather peculiar environmental conditions present – a low temperature, but one that is moderately constant throughout the year, a very high and constant humidity, and a high ultra-violet light intensity due to the altitude and the equatorial position. But we do not really know. In England I have found that these plants grow more actively in the winter. Warmth acts as an inhibiting factor. There is nothing in the behaviour of the particular genera *Lobelia* and *Senecio* in other parts of the world to adduce an inherent character of gigantism in them, rather the opposite when we think of the little blue lobelia cultivated in our English gardens.

It is curious that on the high equatorial Andes a somewhat similar phenomenon appears, with a gigantism again of a Composite genus called *Espeletia*. I have not seen them growing there, but from photographs the similarity of form is close. They are equally difficult to establish in English gardens. A giant *Puya* from there also compares with the more spiky giant lobelias. They are so strange to the ordinary conception of lobelia that some botanists have suggested placing them in a separate genus, and the idea has much to commend it to my mind, in spite of the similarity in floral plan and structure with the more familiar lobelias. There is no plant commonly grown in English gardens with which they can well be compared.

A road, winding and twisting round every little green hill, curling sometimes almost back on itself, led to Fort Portal and to Ruwenzori, the mountains which had always seemed to me by far the most attractive and romantic in Africa. Along this we drove one morning in December.

Fort Portal is the nearest town to the mountain, and here is situated the famous 'Mountains of the Moon' hotel, whose telegraphic address is just 'Romance'. The mountain was invisible.

One of the most exciting events of this day was the proud signpost 'Slow Down Equator'; on one arm was inscribed 'Northern Hemisphere', on the other 'Southern Hemisphere'. We duly slowed down, but did not actually stop. I have since regretted that we did not stop and photograph so unique a signpost.

As night fell we left the cars, pursuing our journey on foot to one of those delightful little rest camps which the beneficent local government of Uganda has sprinkled so freely over the country. There was a second one to receive us the next day, and above that only Ruwenzori, except for the lonely and very beautiful farm where Captain Paul Chapman lived surrounded by mountains and within sound of the incessant music of the Nyamgasani river.

We set off from Chapman's highest camp to the accompaniment of a slow, haunting drum, which beat for several days and several nights for an old man who had just died. We were told that he had been struck by lightning. What an omen and what a farewell sound! The slow, steady rhythm of Africa, so sure and so unhurrying, yet so passionate.

It was all unknown ground. There was no track. We had to cut a path all the way. We soon got used to the routine. We cut generally for two or three days; then all the porters were massed for moving camp. The next day we started cutting again. On some of the lower ridges there is bracken. Everywhere, or very nearly everywhere, in the world there is bracken. Here the cutting was easy, and here we found the first giant lobelia, *L. gibberoa*.

Soon we emerged from the bracken into the forest. Like all transitions of vegetation on Ruwenzori, it was abrupt. The forest covered all the ridges above 6,500 feet and below 10,000 feet and stretched great tongues down the valleys beside the rivers. It was not such a gloomy place as some forests I have seen. In frequent patches a Melastomaceous tree with clusters of large pink flowers, flowers not unlike small single roses, interrupted the continuous green. Few could fail to be thrilled by the giant tree ferns and the wild bananas, some of the most graceful and beautiful of plants. Surely foliage is just as important, from the point of view of decoration, as flower, and these plants gave dignity and distinction to the undergrowth. Although small beside the gigantic specimens of New Zealand, some of the tree ferns were fifteen feet in height. They had enormous fronds emerging from a stem, slender and often charmingly curved, yet so prickly as to repel most painfully any close contact.

The wild bananas bore no fruit, but had hard seed like large rounded beans which germinated freely. This is the only banana, as far as I know, which can be grown from seed. We have had them growing in England from this seed, and their leaves grew eight feet high after two years. In fact we had to part with the largest plant and keep the others root restricted, so that their leaves formed a mere comfortable four to five feet in length. One friend brutally described them as the bananas which 'have never done anything about it', referring to the absence of edible fruit. If, however, we consider their biological purpose as growing and reproducing, the other bananas with fruit but no seed would rather merit this description. The vast leaves of bright emerald green were edged with pale pink through which the sunlight penetrated in a translucent glimmer. The midrib also was pink. I know no leaves through which sunlight penetrates more beautifully than those of the banana. Every vein was delineated like the barbs of a fine quill feather or the crests of the waves of the sea. To stand underneath them and look up at the sun provided a real thrill. The leaves were largely untorn, and did not present the tattered appearance of the common banana. Although often ten or more feet in height, the plants were practically trunkless, the leaves arising from a great cradle formed by the old leaf bases, to which age had imparted a deep crimson colour. Beside the small plant presses these leaves indeed presented a botanist's dilemma. We found many such dilemmas on the mountain.

A group of these wild bananas and tree ferns grew beside a delightful little stream, and at the ford the sunlight penetrated and lit up the out-stretched fronds of the tree ferns and the big leaves of the wild bananas with peculiar brilliance. At the edge of the stream grew a very beautiful pink balsam. Luckily, all the rivers were low and we were easily able to reach this plant. Its flowers seemed to float on the end of slender red stems like delicate shell-pink butterflies with wings outspread. It was certainly one of the most attractive balsams I have ever seen. This plant I introduced to England, and in a cool greenhouse it flowered continuously for a whole year, but was lost during the war years. We had not reached the zones of frost, and so the plants here cannot be absolutely hardy in England.

From Mt Elgon we brought seed of an even larger white balsam, *Impatiens elegantissima*, with flowers several inches across, hovering on a long pedicel like a giant white butterfly, marked with deep crimson and with a spur several inches long. It is a very unusual and fine flower. It is still with us, although from a later importation, and makes tubers like a dahlia which have survived the winter outside in my present Sussex garden. It grows up again then to five or six feet, flowering in late summer until the frosts cuts it down. It is easily propagated also from cuttings and so is worth overwintering in this way as an insurance against a very hard winter. In the greenhouse it proved too large.

The growth everywhere was very luxuriant; the trees were covered with tangled lianas, festooned with streamers of liverworts and tussocked with moss. This was, indeed, a place in which to stand and ponder, one of those rare corners which seem outside the ordinary busy world, a place where the clock stands still and perfect inward peace blends with an outer peace. Such places and moods are rare. This one will be a memory to us for long.

A forest of vast bamboos formed the next zone. Above us the feathery spikes formed arches over our camp, and the tall straight stems helped the feeling that we were in some ancient cathedral. Only a dim and fitful light penetrated. Our camp fires made from dead bamboos flickered like small candles against the overpowering atmosphere of the forest – and a very real forest these bamboos formed. It was a curious green world: the roof was green, the stems of the bamboos were green, the ground was covered with ferns and mosses in innumerable shades of green. Only an occasional dead and leaning bamboo gave a touch of purple or brown. On the bamboo stems were huge purple slugs. There were few exciting plants in this zone, with the exception of the fine scarlet Amaryllid *Choananthus cyrtanthiflorus*, a monotypic genus, peculiar to Ruwenzori. It has pendulous flowers hung like a mop round the top of an eighteen-inch stem. It has flowered in England and been given an Award of Merit by the Royal Horticultural Society. There was also a wicked-looking green, white and chocolate *Arisaema*.

Although we had cleared quite a large space for the camp, before we left the bamboos had arched again over us, so flexible are their stems. Below them our men flitted silently about like shades and shadows from the underworld. There was little sound, only a rustle of the leaflets, a soft murmuring in the wind and an occasional ghostly creaking and cracking as of some spirit laughing at us. This almost enchanted world seemed to have a deadening and depressing effect on the camp. Few men sang or shouted or laughed. They ate their food in silence whispering in muffled tones. Through the forest bats flitted silently – bats in the bamboos.

Gradually the bamboos diminished in size, dwindling from fifty feet to fifteen, until suddenly we emerged into a zone of tree heathers. Imagine a haunted wood composed of ordinary ling heather magnified fifty times; there were trees fifty feet high instead of bushes of one foot, twisted into weird shapes and gnarled so that each resembled a drawing by Arthur Rackham. Out of each trunk glared a face, sometimes benign, more often wicked, and bearded with streamers of lichens and mosses.

Looking out between the tree heathers we could see far into the mountain across a wide expanse of ridges and gorges, but there were no signs of either the lakes or the snowfields which we had hoped to see. It seemed a wild and desolate expanse, in very truth a place where no man lived or would be likely to live. It was not only bare, but mysterious and unearthly. There was no sound. Here the silence became the voice. When grey and misty, it seemed to present a challenge to the man who invaded its solitude; the mountain appeared antagonistic to man and tried to frighten him back again with its uncanny aspects, its cold, its dampness, and the rather putrescent smell which arose from the *Mimulopsis*. Even from the highest branches dangled long sulphurous yellow strands of the *Usnea* lichen, the old man's beard of many travellers, which one of my companions declared reminded him of the hair of Boticelli's angels, an extremely apt comparison in the sunlight. In the mist they resembled nothing so essentially happy, but appeared rather as some melancholy ghosts, not of the animals, but of the vegetable world – the lost souls of a past vegetable glory flapping their branches and stretching out to frighten the wretched man who dared to penetrate such places. It is, indeed, a place of mystery, haunting when these shapes stand out dimly from a background of swirling mists. Among these the stiff spikes of the lobelias barred our path like figures with upright lances. In present-day life such plants seem out of place: they are rather the complement of prehistoric man, or even of the giant reptiles and pterodactyls.

When the sun shone, and it frequently did so for us, the aspect of the mountain changed very quickly and immediately became friendly. Everything then smiled at us; the pink and white everlasting flowers opened into a mass of colour, while the most gorgeous little blue sun-birds appeared and flitted among the lobelias, poking their long beaks into the blue flowers and climbing with agility round the great spikes. The male is indeed resplendent with glossy metallic feathers of brilliant turquoise and emerald, but the female is a dingy brown. They are the African equivalent of the humming-bird.

This heather forest was the zone of the giant *Senoecios* and *Lobelias*. The sheer exuberance of the growth of the giant groundsels and the lobelias is astounding and thrilling. One rosette of *Lobelia bequaertii* would be several feet across, and would have several hundred closely-packed leaves, shining purple and radiating from the centre, where a drop of water would be enshrined like a jewel at the heart of the world. When this lobelia flowered it threw up a stiff green obelisk-like spike, six feet high and nearly a foot in diameter, monstrous and bizarre, but very much in keeping with the surroundings. Between the bracts the deep purple-blue flowers appeared, a good colour, but unfortunately so masked by the stiff bracts that the general effect of the spike was green. The other dominating species was *Lobelia wollastonii*, named after Dr A. F. R. Wollaston who visited the mountain on 1906. Its spike is a glorious powder-blue. When the sun touched the dewdrops on its blue flowers and grey bracts the whole spike seemed to sparkle with a silvery radiance. Among the spikes darted iridescent sunbirds looking like emeralds and sapphires.

Often the plant is twelve to fifteen feet in height and the flower spike is six or eight feet. The bracts are long and woolly, pendulous and densely covered with a greyish-blue pubescence. The flowers emerge between them and are more conspicuous than in most of the other species. The stem is pitted with a decoration of regular diamonds, which makes even the dead plant interesting. It is abundant in the heather forest zone and in the more alpine zone above, extending from 11,000 up to 14,200 feet, where the icicles hung on its leaves and the water from the base of the glacier flowed directly on to it. In the heather forest zone the soil is an inky-black, waxy, somewhat gravelly peat, which is very acid. It is perpetually damp.

Unfortunately this magnificent lobelia is not at all vigorous in growth in England and does not appear to be nearly so easy of cultivation as some of the other species mentioned. It is just possible that there may be difficulties connected with mycorrhiza. On Ruwenzori it has often been photographed half covered with snow and with icicles hanging from the tips of its leaves, but Mr McDouall reported that it was not hardy at Logan, a famous garden in a very mild part of Wigtownshire now maintained as a trust by the Royal Botanic Garden, Edinburgh, nor did we manage to maintain it out of doors in Surrey. Although it often grows in very damp places on Ruwenzori it is not a plant of the peat bogs as is *L. bequaertii*, and a plant tried in my little peat bog soon died.

We had a camp at about 1,200 feet in a most lovely valley. My companion, Stuart Somerville the artist, called it 'Paradise Valley'. It was walled like a natural garden, with grey hills covered with everlasting flowers, heathers and tree grounsels. It seemed different from the other parts of the mountain. A fairy-like but very kindly spirit seemed to pervade it as the little blue sun-birds flitted in the sunlight. They were more numerous here than anywhere else on the mountain, and the air was full of their twittering-tweeting. Everywhere there were flowers, bushes of white and pink everlasting, powdery-blue lobelia spikes, the purple of the *Lobelia bequaertii*, and the golden of the tree grounsels, yet it was an orderly, not a tangled riot as on the rest of the mountain. It was like a place imagined in dreams. We were very happy as we rested there before breasting the steep slope up to the pass. Around us the porters shed their loads and rolled about the grass and heather, chatting easily. These rests on the mountain march were some of our pleasantest times. By now they found little suitable material for smoking, and were delighted with the cigarettes which we distributed. When everyone had smoked a bit, the Nyampara and I would go round rousing them, 'Kwenda, Kwenda, tugende', lifting the loads back on to their heads.

I saw my first giant senecio on Mt Elgon, but they are very similar in general appearance on both mountains. All the way from England I had wondered about these plants. The reality, however, surpassed my expectation. There he stood at a twist of the path, where it descended into a dip to cross a small stream by a rickety bridge; a veritable tree over twenty feet high, branched, gaunt, and with a certain pathetic, bizarre and indescribable look of unreality as of an old man, transported from another planet or age and set down to confront the present world. 'Senex', indeed, means an old man, and these trees are veritable 'Old Men of the Mountains'.

The trunks are twisted and contorted often into all manner of weird shapes, so that some become almost more animal than vegetable; they are surmounted by mops of foliage, like great lax cabbages. The leaves are very large, sometimes three feet in length, and of a rather fierce shade of metallic green. The old leaves do not fall, but remain attached to the tree, dangling as a dead, slowly-decaying mass around the trunk below the rosette. Sometimes they are so numerous that the whole trunk becomes a pillar of dead leaves with a central core.

At Bulambuli, one of our camps in the bamboo forest, there was none in flower, but higher up in the alpine moorland zone we found the giant groundsels flowering frantically. From the centre of the cabbage crown would emerge a vast spike, sometimes three or four feet high and branching repeatedly. The flowers of the higher species were very similar to those of the common English groundsel, except for size and number, but those of the lower species were always much more ornamental having long ray florets (petals, to the non-botanist) like the ragwort or yellow garden daisy. Some of these flowers would be an inch and a half in diameter, and one spike would bear a hundred or more, so that the effect was very striking.

At first the sight of the giant groundsels dominated us, and their bizarreness seemed an ever-exciting and thrilling wonder, but after a few days we began to accept them as part of the landscape, to expect their presence rather than their absence – so soon does habit dull the sense of wonder. On each mountain they were slightly different in appearance and each time thrilled us anew when we came upon them after an interval in the plains. There are no canaries on the African mountain, the more the pity. These giant groundsels would feed all the canaries in the world for a time approaching all eternity, but I suspect that they would be pretty tough and would give our poor canaries a bad tummy-ache. Rabbits might cope with the leaves, but would be unable to touch them while growing, owing to the trunk. However, there are no rabbits on the African mountains, although there are a few hares. In fact, there are very few large animals.

With the lobelias, these senecios form the most striking element in the flora of the higher zones of the Equatorial mountains. They are, however, probably less suitable for cultivation in this country, owing to the long gaunt woody stems which most species form, and their growth in England seems much less hardy and vigorous than that of the lobelias. Mr McDouall reported from Logan that they were delicate with him when young, but appeared to get hardy when large. In the higher species the leaves are covered with a dense white tomentum on the under surface, like a silvery fur coat. *Senecio gardneri* from Mt Elgon, 13,000 feet, is a fine example of this.

For horticultural purposes I think that the dwarf *S. brassica* from Mt Kenya or *S. brassicaeformis*, a somewhat similar species from the Aberdare mountains, would be the most valuable. These plants do not form long woody trunks and the rosettes are almost sessile on the ground. The leaves are packed closely together and in *S. brassica* are thickly covered with a close silvery-white indumentum. The flowering spike rises to a height of four or five feet, the ray florets are moderately long and all the stem of the spike and the penduncles are covered with the thick silvery-white hair. For some time we had several plants of these species growing quite vigorously in Surrey, but unfortunately they have now all died. *S. brassica*, like most of the aborescent senecios, is a water-loving plant and is found chiefly in boggy places from 11,000 to 14,500 feet. Another rather dwarf and very attractive species is *S. alticola* from the Virunga mountains, but this does not possess long ray florets like *S. brassica*.

The lobelias from Mt Elgon have also proved easier of cultivation than those of Ruwenzori. *Lobelia elgonensis* I had for some time growing strongly, although it was badly damaged by a very severe winter. At one time the rosettes looked completely dead and limp and I despaired of them. However, they sprouted again out of the middle when the spring came. *L. elgonensis* is very similar indeed to *L. bequaertii* of Ruwenzori, both in flower and appearance, a stiff green obelisk. I was much excited when I first discovered it and gathered a whole plant under my arm like a great baby and with it marched proudly back to camp. It proved a heavy burden. It was of course, impossible to press such a monster whole, so he was cut into thin longitudinal slices.

All these species of lobelia grew in damp places, some actually in bogs and on the banks of streams, but we found on Mt Elgon one species which grew among rocks in drier places right up to 14,000 feet. This was *Lobelia telekii*. It has long hairy bracts, drooping and covering the pale-blue flowers, which gave it the appearance of a gigantic woolly caterpillar petrified and stood on end. It belongs to the same group as *Lobelia wollastonii*, but is not nearly such a fine plant. It appears, however, to be one of the hardiest and easiest of cultivation in English gardens.

On Mt Kenya there were particularly magnificent stands of *Lobelia telekii*, like groups of great petrified woolly bears. The species was named after Count Teleki, a Hungarian explorer. They suggested for us 'family life on Mt Kenya'. There were two in particular which amused us, a tall one and a short one side by side. From a little distance they looked like two figures out for a walk, a tall man and his short wife. Then suddenly they came over the crest into sight of the snow peaks and had stood there gazing at them petrified with surprise.

Both dwarf creeping hypericums (St John's wort) with little flowers, and shrubby almost tree-like hypericums with large flowers, are found on these mountains. It is the latter which I would like to see introduced into English gardens. In the commonest species, *Hypericum lanceolatum* from Mt Elgon, the flowers are about the same size as those of the commonly grown *H. calycinum*, the stigma and ovary are a handsome crimson, while the flowers are borne abundantly and the plant grows to the size of a small tree, often twenty or more feet in height. The foliage is evergreen and very similar to that of the commonly cultivated white veronica. But in my opinion the finest hypericum of all, the finest hypericum I have seen anywhere, is *H. bequaertii* from the heather zones of Ruwenzori. This plant, also, attains the size of a small tree and bears its flowers pendent at the ends of short branches. Unlike the other members of the genus these do not open flat, but are cup-shaped like small tulips hanging from the branches. Unfortunately these species have not so far proved very vigorous with us, although some grew to shrub size in the late Mr Norman Hadden's jungle of rare plants at Porlock in Somerset, while they have also grown in some Irish gardens, and I heard of one in Cornwall before the war that was between five and six feet in height but had not then flowered. I do not unfortunately know of its present state and my informant has died.

There are several species of *Kniphofia* on these mountains and although none of them is as showy as the commonly cultivated hybrids, many of them are attractive plants and well worth cultivation. One of them, *Kniphofia snowdenii* from Mt Elgon, is in cultivation in England and is now to be found in a few nurserymen's catalogues. It seems to be generally hardy. The flowers are yellow and are borne in a loose raceme, two or three feet in height. The plant somewhat resembles a giant *Lachenalia*, and curiously the cultivated form is genetically a sterile triploid, but like other kniphofias it spreads by stolons.

Epiphytes did not for the most part prove interesting here and the epiphytic orchids were extremely dowdy, being far eclipsed by the terrestrial species. There was, however, one epiphyte which greatly attracted me, *Canarina eminii*.

The genus *Canarina* belongs to the order Campanulaceae and only one species, *Canarina campanulata* from the Canary Islands, is at present generally cultivated in this country. *C. eminii* grows as an epiphyte on Mt Elgon, between 6,000 and 9,000 feet. The branches and leaves are glaucous and pendulous. At the ends are large pale orange bell-shaped flowers lined with deep crimson. Inside the bell is a large club-shaped stigma like a clapper. This plant is easily cultivated in a cool greenhouse in this country and forms a permanent fleshy tuberous root. After flowering it quickly dies down, and when we used to grow it before the war from my collected tubers we used to dry it off like a dahlia in the winter and it quickly started into growth early the next spring. I gave a tuber to the late Mr E. A. Bowles and he quickly propagated it and used it in urns in the garden in the summer where it hung down gracefully over the edges. Now, alas, it has become very rare in England, probably only existing still at the Cambridge Botanic Garden. *C. eminii* has a naturally graceful habit of growth, which combines with the beautiful flowers and the ease of cultivation to make it a valuable plant. During the summer it can be planted outside. The Royal Horticultural Society gave this plant an Award of Merit when I showed it at the Chelsea show in 1938. I grow here, in a cool greenhouse, *Canarina campanulata* of which I was given seed from a famous Riviera garden where it grew outside. Here it flowers in March and April, starting into growth too early for successful growth outside all the year round, and the bells are not so long nor so deep in colour.

There are several umbellifers which we thought attractive, in particular *Peucadanum kerstenii*, an almost arborescent species from the heather forest of Ruwenzori with very fern-like foliage, and *Heracleum elgonense* with creamy flowers from the higher zone of Mt Elgon. Mr K. McDouall used to have *Peucadanum kerstenii* growing at Logan but reported that it was not a success and I have not heard of it anywhere in cultivation in recent years.

There are no gentians on these mountains, but the large-flowered dwarf white swertia, which we found on the Aberdares, seemed to me a most attractive plant and I regret that I have not got it growing. Both on Mt Kenya and Mt Elgon we found *Delphinium macrocentron* and I do not know of any other plant which attains quite the same subtle and electric shade of blue: the nearest, perhaps, is the little *Corydalis cashmeriana*. It seems to combine the clearness and etherealness of the sky at sunset with a touch of green, and also the more vivid, more chemical tone of a solution of copper sulphate. It is distant and elusive in one glance, yet near, electric and vivid in another.

From Ruwenzori the rampant creeper *Thunbergianthus*, with its big pink trumpet-shaped flowers, would be desirable for the large cool greenhouse or the orangery, but none of the seeds we collected has germinated and other attempts to introduce it have failed. There is also a large golden-flowered *Sedum*, at about 11,000 feet, on Ruwenzori, which I would like to have been able to introduce. Near the foot of several of the mountains we also found plants such as *Gloriosa superba* and *G. virescens* and *Haemanthus multiflorus*, but these are not elements in the mountain flora and are known in this country.

Although the chief plants of these mountains are by now known, I am sure that there are many which could be profitably introduced into English gardens. There is a possibility also that the second and third generations raised from seed entirely grown in England might prove more hardy and better adapted to English conditions than the first generation. The scenery and the colour of Ruwenzori stand out in my memory as the greatest event of the whole African trip. Nothing on any of the other mountains we visited could approach it. It seemed an enchanted world, unreal as if we had suddenly stepped into a fairy pantomime staged by some very mad genius. All was mysterious, while in the sun the whole landscape had an intensity of colour never found in England.

Just thirty years later I visited the Golden Gate Park in San Francisco, a very mild and protected area situated close to the Pacific Ocean and apparently frost free. Here the giant lobelias were growing better than I had seen them in any other garden, and there were great rosettes of several of them which brought back pleasant memories of Africa. All this area has a good deal of sea fog so that the atmosphere is moist for much of the time; and the length of day and night is more even throughout the year and nearer to that which they experience in East Africa. The rhododendrons from Java and New Guinea also grew well outside and probably those from Borneo would have done so too.

On Mt Elgon, particularly in the alpine and subalpine grassland, there are numerous small bulbous species belonging to the genera *Crinum*, *Romulea, Hesperantha, Gladiolus, Oenostachys*, some of which I had in cultivation in a cool greenhouse and which seem desirable. *Crinum johnstonii* has bulbs the size of a football, long strap-shaped leaves, and a fine large flower spike about three feet in height which bears several large white flowers, prominently streaked with mauve down the centre of each petal. The brilliant scarlet-flowered form of *Gladiolus quartinianus*, from the alpine zone of the Aberdares and Mt Kenya, is also a most attractive plant and very rich in colour. *Oenostachys dichroa* is a curious plant from Elgon. Its corms and habit of growth are similar to those of a gladiolus. The flowers are small and are shielded by large bracts which are generally a deep mauve. The late Sir Frederick Stern flowered this plant at Highdown in Sussex but reported that the bracts were green and showed no trace of mauve coloration. It seems as if this might be a physiological character rather than a purely systematic one.

The other great floral delight of Uganda lay in the blue and white water lilies of the lakes, especially the shallow ones such as Lake Kioga.

These were fringed with the graceful papyrus. The lake is glorious and we could see down to the bottom, where little fish play among the water weeds. It was wholly attractive in colour and fairylike in form. Under the water-lily leaves there are very dark shadows contrasting with the green shininess of the leaves, over which drops of water run like quicksilver, splitting and joining in a merry antic, and for ever trying to find their way back to the water below. The blue flowers are not borne nestling in the water as in the English species. They stand proudly six or eight inches above it. Among them are a few pink flowers and some big white ones, probably the true *Nymphaea lotus*, and maybe the Egyptian lotus sacred to Isis. Beautiful though it is, this plant cannot compare with the sacred bean, *Nelumbo nucifera*, the lotus figured and revered by the Chinese; a superb plant, without equal in the whole plant world. It also grows in marshes and even out of mud. In the black mud of the lake these blue water lilies conceal long fat tubers, which Africans sometimes pull up and eat. This is not, though, the source of the lotus eaters of Greek mythology. This lotus is yet another plant, a shrub named *Zisyphus*. The water lilies form a fringe twenty to fifty yards wide all round the edge of the lake and around every island, while in shallow parts there are acres, and even square miles – veritable seas – covered with them. As the sun comes out all the flowers open. When the sun goes in all the flowers close. The scent is delicious, and the silence is broken only by the slight buzz of the bees among the flowers and the whistling of the breeze which always seems to sweep over the lake, a delicate sound, like the whispering murmurs of the sea in a shell held up to the ear. Among the water lilies vast pelicans floated majestically along, while black and white kingfishers darted over-head and statuesque storks and herons stood immobile waiting for their fish.

CHAPTER EIGHT

North America: In Search of Lilies

The visit which I describe lasted only six weeks and nearly two of these were taken up with meetings, but it is surprising what a number of interesting plants it is possible to see and what a great range of country one can visit in four weeks, particularly with the aid of kind and knowledgeable friends who know where to take one. The North American mountain plants are not so well known to our gardeners as the Himalayan or Middle Eastern plants, but it is a very rich flora with many exciting members which should be tried more in English gardens.

In July I flew to Boston and then attended the show and annual meeting of the North American Lily Society held in Worcester, Massachusetts, about an hour's drive inland. There were many wonderful lilies at the show, including many new spotless hybrids which I had not seen before, and it was good to meet so many real lily enthusiasts. But one of the top prizes went to a magnificent spike of the yellow form of *Lilium canadense* which grew wild within a few miles of Worcester, and it was a great thrill to see it there. It is one of the most graceful of lilies, the deep yellow flowers floating on their long and curved swan-like stems like large yellow butterflies. Always it grew in very damp places at the edge of small streams, and undoubtedly it will never succeed under very dry conditions.

It was evening and the sun was low and lit up the petals to a rich pure yellow, the colour of the yellowest butter. A large spike looks like a deep yellow candelabrum, and the flowers, each several inches across and on a long stem, are well balanced in proportion to the stem.

Without the help of knowledgeable guides like Dr George Slate, the editor of the Year Book of the North American Lily Society, and later in the west, Frank Ford, who has travelled all over the west and has amassed over the years a vast hoard of unique knowledge of the Pacific coast lilies and exactly where they grow, I should perhaps not have seen half of those I did see in the time, perhaps even none at all for they are very local, but they are as exciting and, I think, as beautiful as the Sino-Himalayan lilies. North America is wonderfully rich in her native lilies. I think there are at least twenty-one described and generally accepted species, and are several records of plants which might on further investigation be described as new species. It is sometimes difficult to tell the difference in the wild between natural hybrids, which are not yet a population, and established but rare species. However, in that way new species occur.

On this trip I was also interested in seeing as many of the cultivated hybrids as possible, and as well as the North American Lily show I was able to see the trials of hybrid lilies near the Arnold Arboretum, and to visit the Oregon Bulb Farms which were certainly a revelation in lily growing, as well as several other lily nurseries near the west coast.

From Worcester, Dr George Slate kindly drove me two hundred miles north-westwards to his home at Geneva, nearly as far north as Lake Ontario. Here, he was able to show me an outstandingly lovely crimson-scarlet form of *L. canadense*, again growing only in moist places. This was presumably the var. *coccineum*. I never saw yellow and red forms growing together, although Mr Wyatt did so the year before, farther south in Connecticut. Professor Slate writes that in general the yellow form may be east of the Hudson River and the red west of the river. I do not know the distinguishing characters of var. *coccineum* and var. *rubrum*, but a plant with darker red flowers growing at Wisley has always been known, and probably rightly, as var. *rubrum*.

These are the forms of the eastern lowland districts, east of the great Appalachain chain; but growing actually on the mountains farther south, from Pennsylvania to Alabama, is var. *editorum*, also with scarlet-red flowers and distinguished by its broader and less tapering leaves. It also has another great difference in that it is reported as growing in much drier places, on rocky wooded slopes and open mountain meadows. I have never seen it and would much like confirmation of this.

A number of other forms have also been named and sound well worth every effort to find and introduce again, There is 'Redwing' a form of var. *editorum*, and 'Golden Rule', a particularly beautiful unspotted light yellow form collected by the late Mrs Norman Henry in eastern Pennsylvania. She also reported another form from the mountains of east Virginia with smaller deep red flowers which sounds very desirable.

L. canadense is the prior name in the group which includes *L. michiganense*, *L. superbum*, *L. iridollae* and *L. michauxii*, and obviously these are all closely related. All have flowers in varying shades of orange and yellow and red with recurved petals, rhizomatous or stoloniform bulbs and whorled leaves, and all grow in moist places.

We also saw *Lilium philadelphicum* which is a very rare and difficult lily in cultivation. It was growing in absolutely the opposite kind of conditions to those of *L. canadense*, on a dry sunny hill-top with open vegetation among small shrubs, and in sandy soil almost dust-dry near the surface. It is a lily with upright flowers and petals narrow and claw-like at the base so that they look deeply divided, a very strong reddish-orange in colour, heavily spotted with deep maroon. It is said to be variable in colour but we only saw it in this one place. The spikes were short, usually not over two feet and with only one or two flowers, but larger ones are known. In cultivation in the country it is difficult and short lived and appears to be intolerant of winter wet, so that a glass over it then is probably a help. Some even suggest that it should be kept dry from the time of flowering in early July on through the summer as well as in the winter, for much of which it would normally be under snow.

Fron New York I was able to fly right across the country several thousand miles to Seattle, close to the Pacific coast and in the State of Washington, a hop no native lily has been able to make. Here, Brian Mulligan, head of the University of Washington Arboretum and an old Wisley man, and his wife kindly looked after me and took me up into the mountains to see some of their native flora. I also saw several very interesting gardens, but it is of the wild flowers that I will write now. Even beside the highways we came on a big stand of *L. columbianum*, perhaps fifty spikes in quite a small area. They were coming up through a carpet of *Linnea borealis*, the pale pink American form, which was also in flower and made an exciting combination. The Turk's-cap yellow flowers are small in comparison with some other lilies, but when seen in a large group it seems well worth growing, and it is not reported to be a difficult lily although it is rarely seen in gardens here.

It was a week-end, and the first day we headed for the Mount Rainier National Park, 377 square miles of mountain territory rising to the permanent snows of the 14,410 foot Mt Rainier. These National Parks are very popular and thousands throng there for the week-end. There are good roads through and numerous camping sites, also good trails where one can get away from most of the crowds which throng the park information centres. Flowers and animals are strictly protected and there is an organization of park rangers. The result is that many of the flowers grow in great abundance. I wish we used and respected our National Parks as the Americans use and enjoy theirs. The lower slopes are covered with magnificent stands of conifers: Douglas firs, *Abies grandis*, Western Hemlock, *Tsuga heterophylla* and several others. So many of our best forestry trees come from this area and were introduced by David Douglas through the auspices of the Royal Horticultural Society just one hundred and forty years ago. Near the Pacific coast, Douglas firs reach three hundred feet or even more in height. Another common tree was the Madrono, *Arbutus menziesii*, perhaps the finest species of this genus with its beautiful reddish-mahogany peeling bark and hanging clusters of strawberry fruits in the autumn and winter, while the flowers in spring are like great clusters of lily of the valley or pieris, though without their scent.

When we got up to the zone where the snow had only recently melted, clusters of creamy-white *Pulsatilla occidentalis* appeared, as beautiful perhaps as the more familiar *P. vernalis* of the Alps, great gleaming egg-shaped flowers with golden centres. The greatest thrill, though, lay in the erythroniums, the golden-yellow *Erythronium grandiflorum*, the Glacier lily, which grew in myriads beside the melting patches of snow, just as solda-nellas grow in the Alps. The flowers were larger than those of the dog-tooth violets we grow here, mostly single, on six-inch stems, but occasion-ally two, poised delicately and reflexed at the margins of the petals as a miniature of a lily. I only wish that it would settle in English gardens and spread as it did in Oregon since it was a most lovely sight. But I know of no garden in which it has done so, although the paler yellow *E. californicum* and the creamy hybrid 'White Beauty' as well as the pink *E. revolutum*, both from Oregon and California, seem to settle down much better, but they do not come from such a high altitude. I suppose that we saw the yellow *E. grandiflorum* at about 9,000–10,000 feet, but it was late July and they would have flowered earlier a bit lower down. At any rate they are one of my greatest flower memories. I was not able to see the white *E. montanum*, the Avalanche lily, which I believe grows in equal masses on other parts of the mountain and also on Mt Hood, and is reported to be equally difficult in cultivation even in other parts of America.

The penstemons and the phlox were other mountain flowers of value for rock gardeners. *Penstemon fruticosus* was the commonest species and grew on shady stony slopes in mats several yards across covered with pinkish-purple flowers – lovely outlined against the snow peaks. This is one of the largest genera of the Pacific coast mountains and Professor Munz in his *Flora of California* gives no fewer than fifty-eight species, so it is not easy for the stranger to recognize many. Later, south in Oregon, we found a white form. *Phlox diffusa* varied much in colour from white to fairly deep mauve but always it was prostrate, growing on rather stony, open slopes and covered in flower. It differs only slightly from the *P. douglasii* whose varieties are so common in our rock gardens and which also comes from this area. Other interesting plants included *Phyllodoce glanduliflora*, whose greenish-yellow bells draped over an old log, while on the most vertical rocks were great bosses of *Cassiope mertensiana* although unfortunately their white bells were finished by July. Locally it is called white heather, which shows the confusion of the use of vernacular names in different parts of the world. One could easily have spent with interest several weeks on the massif of Mt Rainier alone, and probably would have found different plants every day.

The next day we went to another smaller peak in southern Washington State and found it was equally rich in interesting flowers. A particular excitement for me was the sight of *Lewisia tweedyi* in flower, both pink and white forms, fully living up to Farrer's most grandiloquent description. It was growing on shaly slopes where drainage would be excellent and it would have been under snow for much of the winter. It really is a superb flower, about two and a half inches across, regular and opening almost flat, with a pale creamy-yellow base colour flushed with the palest apricot-pink. It grows well in England as an alpine house plant but one rarely sees good specimens outside. Farther south we saw other lewisias, forms of *L. howellii* probably, but they had finished flowering and were growing in much hotter and drier conditions.

Beside the *Lewisia tweedyi* were *Clematis columbiana*, an interesting purple-flowered species resembling the alpine atragene in form of flower and habit; *Aquilegia flavescens*, with yellow and cream flowers, slightly touched with pink; *Dodecatheon pauciflorum* and *D. jeffreyi* the Shooting Stars, and a smaller one with white flowers which has been named as *D. dentatum*. Always they grew in damp places among lush herbage. The *Castillejas* made most brilliant splashes of scarlet and particularly fine was *Castilleja miniata*. The colour is in the bracts. Unfortunately these are root parasites which makes them difficult to establish, but some alpine gardeners have managed to flower them in pans in this country.

The veratrums grew lavishly in open glades and *Veratrum californicum*, with large plumes of greenish-white flowers, seemed the equal of our *V. album*, although perhaps it was slightly denser in its flower head. Another rare and unusual plant of this area was the creamy-white *Rhododendron albiflorum* growing also on steep banks. It is obviously very local and we saw it only in one small area. It was a loose deciduous shrub about four feet tall, and the flowers were slightly pendulous, each about an inch across, rather open bell-shaped and growing in pairs or singly in the axils of the leaves. It, again, has the reputation of being very difficult to grow in cultivation and I do not recollect ever having seen a spray of it shown at one of the R.H.S. Rhododendron Shows where perhaps in a good year one can see the greatest gathering of species of rhododendron from all over the world.

From Seattle I flew down to Portland where Jan de Graaff, founder and till recently head of the Oregon Bulb Farms met me and was my host for several days. I had heard much about his lilies but the reality surpassed all my expectations. The Oregon bulb-fields are widely dispersed and we drove many miles between them. One of the secrets of de Graaff's success in lily culture, I am sure, is his system of shifting cultivation, the seedlings being planted in fresh ground which has not been used for lilies before, thus obviating many of the diseases which plague lilies in our gardens. It surprised me how vigorous and free from disease of all kinds were these stocks, even when grown together in masses as a field crop. It is a very large-scale enterprise. Another factor is the heavy feeding with artificial chemicals, and this was advocated by many lily growers. It seems to me that lily growers in this country could afford to experiment much more generously with the use of fertilizers. The hotter sun that Oregon enjoys as compared to our summers must also be another factor in their success. Only a few of the Oriental hybrids of *auratum*, *speciosum* and *japonicum* were out, but some of the Aurelian strains were magnificent. No other word would be appropriate. 'Golden Splendour' has been derived from the deepest golden flowers of the 'Golden Clarion' strain and gave open trumpets six inches long and as much across. Rather close was the 'Sunburst' strain, but here the flowers were a little more star-shaped and a little less trumpet-shaped, and up to eight inches across.

I walked slowly down the rows to select flowers for photography and it surprised me how uniform they were. It was hard to pick one as being outstanding beyond the others, although all these had been raised from seed. It is obviously much easier to keep vigour and health in such a strain raised from seed than in clones propagated vegetatively from scales. The 'Heart's Desire' strain had beautiful, white, open bowl-shaped flowers with yellow centres and were obviously good garden lilies.

Another area that particularly interested me was planted with patches of many of the species and these had far more variation than we saw in the strains, in particular in such species as *Lilium wardii*. This suggests the possibility of selection of more vigorous and deeper or clearer coloured forms than we are used to.

In the breeding houses we also saw many interesting plants, even the very rare *L. bakerianum* from western China and northern Burma which I feared had died out of cultivation, yet here were a dozen or more in flower. Some of the *auratum-speciosum* forms and hybrids were superb, in particular *auratum* var. *virginale*, a form with white flowers and a pale yellow wave in the centre. It was unusual to see an albino of this type growing with such vigour. 'Pink Glory' had some *japonicum* as well in its parentage, and was a delicate clear pink bowl-shaped trumpet only just faintly spotted at the base. 'White Tiger' was a creamy-yellow deepening towards the centre, with quite large horizontal flowers and slightly recurved petals, which promised to be the beginning of a new and valuable

F

strain. It is only in recent years that we have begun to realize what a valuable parent *L. tigrinum* can be. *L. brownii australe* was so tall that I needed a ladder to photograph it, a lovely white trumpet, but, coming from Hong Kong, it is probably too tender for outside cultivation here.

The seedling houses were another revelation, many thousands of seedlings growing thickly and healthily in long trays. Very little is left to chance here and one got the impression that lily cultivation had really been mastered at one place in the world, if not elsewhere. I had the pleasure of a talk also with the late Mr Earl Hornback who was then a sick man and, in fact, died within a few months of my visit. He had been responsible, with Jan de Graaff, for many of the breeding programmes and he showed me some of the meticulous records which they kept. Mr de Graaff also took me to meet other lily growers, among them Mr Edgar Kline and Mr Carleton Yerex, and it was interesting to see their plantings also. For the warm hospitality and their lovely house looking out over a deep gorge of the Sandy river, as well as their superb lilies, I shall long remember my visit to Mr and Mrs Jan de Graaff and I only wish we could grow their lilies better and with the certainty of success that they seem to have achieved.

From Portland I went on to Eugene in south Oregon where Dr and Mrs Milton Walker were my hosts. Here everything was geared to the genus *Rhododendron* instead of *Lilium*, but it was equally interesting although I was too late for the main mass of flowering. Dr Walker has been instrumental in setting up the species foundation of the American Rhododendron Society and had obtained many scions of the outstanding forms of species from this country for the Foundation. Here it is necessary to water heavily during the summer and I envied Dr Walker's elaborate system of sprinklers controlled by time-clocks. Each part of the garden was watered once in three nights during the summer.

Dr and Mrs Walker are also enthusiasts for the wild flowers of their mountains and took me up into several parts of the Cascade range and also on to Mt Ashland. We went up again through magnificent stands of conifers, in clearings carpeted with little clintonias and *Cornus canadensis* in flower. Soon after we came out into the open we began to find, always on sloping ground, *Lilium washingtonianum purpurascens* in flower and this seemed fairly plentiful in this part. It is an interesting lily in that the bowl-shaped flowers open pure white and then gradually darken through pink to purple before dying off, and one can usually see all colours on the same head. It is also a very beautiful lily; although very rare in cultivation in this country at present, one group, given by Mr Wyatt, did flower well that summer on Battleston Hill at Wisley. On this mountain also I saw my only *Calochortus* in flower, a beautiful cool creamy-yellow, probably *Calochortus lobbii*. Here also were more brilliant *Castillejas*, scarlet *Gilia aggregata*, the sky-rocket gilia which coloured great stretches of the mountain, and various lupins and penstemons. *Lupinus lyallii* was lovely with silvery leaves and blue flowers, almost prostrate on the ground. The finest penstemon was probably *Penstemon rupicola* which crept over rocks near the summit, forming mats of strong purplish-crimson flowers, varying in colour a little with the exposure. We also saw more *P. fruticosus* and even a white form, and I spent some time trying to photograph the little humming birds which darted among the flowers. The eriogonums are another feature of these mountains and some, such as the buff-yellow *Eriogonum umbellatum*, would be well worth trying in this country as rock-garden plants.

Many of the mountains in this region are extinct volcanoes and some-times the crater has become filled with a lake to which the volcanic rocks have given a most brilliant colour. We visited the famous Crater Lake in south Oregon and here the water was the deepest blue I have ever seen. The lake is said to be the deepest in the U.S.A. and this may partly account for the wonderful colour in the brilliant sun. It is a National Park and the flowers again were plentiful, lupins and phlox predominating.

We were joined at Ashland by Mr Frank Ford, another real lily enthu-siast and one who has a unique knowledge of the locations of the native species of the Pacific coast; with him I travelled for the next ten days or so, going southwards to southern California by road and crossing between the mountains and the coast. First we went down through the redwoods, a great experience often described, to the Pacific coast, where Oregon and California join. The resemblance of the trees to the pillars of a great cathedral is valid, and the forest floor has the same quiet and coolness as contrasted with the heat of other areas. We next visited Mr Leslie Wood-riff whose lily farm is within sight of the ocean. He has raised some very interesting lilies and believes in the technique of applying the pollen from several different lilies to each suitable stigma, so records are difficult. He raised the interesting 'Black Beauty', the first cross involving *Lilium henryi* and *L. speciosum*, and we were able to see there what a strong-growing lily it is. I particularly liked also some of his aurelians. His family kindly took us as well to see *L. occidentale* growing in a marsh a few miles away, not such fine spikes as the one recently shown at the R.H.S. from Quarry Wood, but still several feet in height with five or six recurved crimson-red flowers to a stem. The marsh was quite moist in places, though the lilies did not seem to grow in the dampest places but among the tussocks. The flowers were a rich deep crimson but had shining green centres with orange spots and the petals recurred gracefully. They were not very large, about three inches across, and had long curving pedicels like a lamp bracket. All this part of California and southern Oregon, however, is more green and moist than the interior because of the coastal mist and fog which rolls in from the Pacific during part of the summer. Inland in southern Oregon we also saw some of the smaller forms of *L. pardalinum* growing beside streams and in dampish places, probably those which have sometimes been distinguished as *L. wigginsii*, but it is really a large and very variable aggregate species.

From southern Oregon we went inland again and southwards to Mt Shasta and Mt Lassen. Mt Shasta is the home of *L. shastense*, another variant of the *pardalinum* group, with smallish yellow flowers heavily flushed with orange. This we found near the foot of the mountain growing in a marsh by the roadside among tall herbage. Higher up the mountain, after a considerable search among a thick scrub of *Ceanothus velutinus*, of which a few greyish-white flowers still survived, and *Arctostaphylos*, we found a spike of *L. washingtonianum minor*. This differs from the *purpurascens* variety in having no basic pink or purple colour in the petals and being very much more heavily spotted. It has also a much narrower trumpet. These lilies have no rain from the time they flower until the late autumn, and perhaps this is one of the important factors in attempts to grow these dry-country species. *L. humboldtii* and its forms and *L. bolanderi* probably come into the same group for cultivation. On the higher areas of the mountain, however, there was still water in the streams from the melting snows above and here one found beautiful yellow mimulus such as *Mimulus primuloides*, penstemons such as the sky-blue *Penstemon heterophyllus* and, on drier rocky ground, the pink pussy-paws *Spraguea umbellata* and a rather prickly-foliaged white phlox.

From Mt Shasta and Mt Lassen we went south-east to Lake Tahoe, near the border of California and Nevada. This is a very popular resort and a most beautiful area with water of the brightest blue with just a touch of emerald in it. On a small cliff by the road we found great bosses of the scarlet *Zauschneria californica* hanging down from the rocks and it was interesting to see the vertical clefts in which it grew, with the flowers in full sun. Nearby, alongside a small lake which Frank Ford knew, we found *L. parvum*, an orange lily and a much finer flower than its name would suggest. The flowers were shaped like large thimbles, semi-pendulous and spotted inside. Probably the form we found was that known as *crocatum*. This was also a lily of damp places although not of a marsh and grew among alder scrub and in semi-shade.

From Lake Tahoe we went westwards into San Francisco, stopping a night on the way outside the city near Sacramento with Mr Lester Hannibal and his wife. He is an enthusiast for *Amaryllis* and *Crinum*, and although the former were not yet out we saw some very interesting species and hybrid crinums. He also showed us an area which in spring had been thick with *Calochortus* and *Brodiaeas;* now it was bare and baked so hard that a pickaxe, stoutly swung from the shoulder, was necessary to collect a few bulbs.

In San Francisco we had dinner with Dr Albert Vollmer and his wife. He is one of the greatest lily enthusiasts of the Pacific coast species and had many interesting points to tell us. Fortunately he has at any rate recorded some of his knowledge in articles in previous issues of the R.H.S. Lily Year Books. We also had an afternoon in the Strybing Arboretum of the Golden Gate Park under the guidance of Dr Jock Brydon, then Director. Here conditions for growth are most favourable. I was fascinated to see the compact growth of tender *Maddenii* rhododendrons here and was told they flowered equally well. Some of the New Guinea species were also growing well and were exciting. Another unexpected sight was the great rosettes of some of the giant lobelias from the East African Equatorial mountains which I had last seen in their native home nearly thirty years before.

A long day's drive took us down to Pasadena by Los Angeles, where Frank Ford lives, and on the way we paused at Santa Barbara to see a little of the Botanic Garden, one of the two which specializes in growing native Californian plants. Here we saw the huge *Eriogonum giganteum*, six feet or more in height, with vast panicles of creamy-white flowers. Santa Barbara must be a most favourable area horticulturally and the gardens of the houses we passed looked better than any of those we saw in other areas.

There was one more lily yet to seek, *L. parryi*, one of the most beautiful of all. This grows in the mountains east of Los Angeles, but unfortunately it had finished its flowering for the year. These mountains, the Sierra Madre and the San Bernardino, are much drier than those farther north and one went up through xerophytic prickly scrub and great grey rosettes of *Yucca whipplei*. The rosettes were often three or four feet across and each leaf was needle-sharp in its point. The flowers were unfortunately over but the spikes still remained, many about eight feet high. Their white bells must be magnificent flowers. Before a ban was imposed on cutting them the species was in danger since visitors, with thick gloves, used to cut spikes and take them back to the city on the roof racks of their cars. This prevented seeding and the main rosette dies after flowering. The lilies only grow in the deep gulleys where there is moisture and often a little stream, and some tree cover has been able to grow. We did not see *L.*

parryi itself, which grows around Big Bear Lake, but we did find capsules of *L. parryi* var. *kessleri* and remains of quite substantial spikes. The flowers are a clear yellow, strongly scented, trumpet-shaped but opening wider than in the type. This is the country of the big-coned pines *Pinus ponderosa* and *P. coulteri*, and magnificent trees some of them were, particularly the former.

I was also able to visit the Rancho Santa Ana Botanic Gardens at Claremont, the Los Angeles State and County Arboretum and the Huntingdon Botanic Gardens at Pasadena all of which I found very varied and intensely interesting. The cactus garden at Pasadena, which shows them all growing among dark volcanic rock, is most effective; this is a frost-free area and *Oleander* and *Lagerstroemia* were in full flower everywhere while the avocado pears hung in heavy green clusters on the trees.

All too soon it was necessary to return eastwards for the Horticultural Congress, and since there was an air strike I travelled by train a great double-decked silver-painted monster which enabled me to see much more of the country. Sometimes it ambled through the mountains so slowly that I felt tempted to get out to see the flowers more closely but would never have caught it up again. On the way I was able to visit the Hunt Botanical Library at Pittsburgh and to see some of their collection of treasures – a great treat.

One of the excursions from the Maryland Congress was a visit to the Longwood Gardens. Some of the old trees aroused great interest in the party, but the evening visit to the Conservatories with their tree ferns all lit up was the special feature and the night-blooming tropical water lilies were incredibly lovely, each plant being lighted over the top. Many had been raised by Mr George Pring of Missouri who was with us. To finish the evening we all sat on the terrace and the fountains played before us, the jets being varied in height and colour and pattern, making a most dramatic and beautiful display against the dark sky.

CHAPTER NINE

A Trek in the Nepal Himalaya

For many years I had longed to visit the Himalaya and to see for myself some of our garden flowers growing in the wild there. From my under-graduate days I had read books by the great plant explorers, Farrer and Kingdon-Ward, as well as accounts of other expeditions. The pattern of plant collectors has changed and few are now able to spend a whole year or even several years as Forrest sometimes did in the field, noting desirable plants in the spring and returning in the autumn to collect seeds or sending native collectors for them. This resulted in a vast spate of new plants from the rich Sino-Himalayan regions, especially rhododendrons and primulas, and patrons of these collectors had to extend their gardens out into their woods to find places for the great numbers of plants that they raised. Thus the woodland garden, one of the most delightful and labour-saving forms of present day gardening, originated. These were times of cheap labour and materials in the garden, and the owners of some of the larger estates could afford to raise vast quantities of plants and to distribute some among their friends. It was of great benefit to horti-culture, and gradually a small proportion of these plants found their way into the nurseryman's list. The majority of the rhododendrons, magnolias and other shrubs which they collected have survived but, alas, only a small proportion of the primulas.

Unfortunately now the Sino-Himalayan regions of western China, northern Burma, northern Assam and south-east Tibet, which have yielded us so many wonderful plants, are closed to Western travellers, but towards its western end Nepal, which was formerly a closed country, is now open and a new style of travel can take one there – a quick flight out through Delhi and Khatmandu, a short flight to Pokhara in the centre of Nepal and a trek which means a long mountain walk, camping at night with porters to carry loads and a Sherpa or two to act as guides and generally control the details of organization. This relieves the trekkers of some troubles and leaves them free to enjoy the country and the flowers. Several firms based in Khatmandu organize all the details, including trek

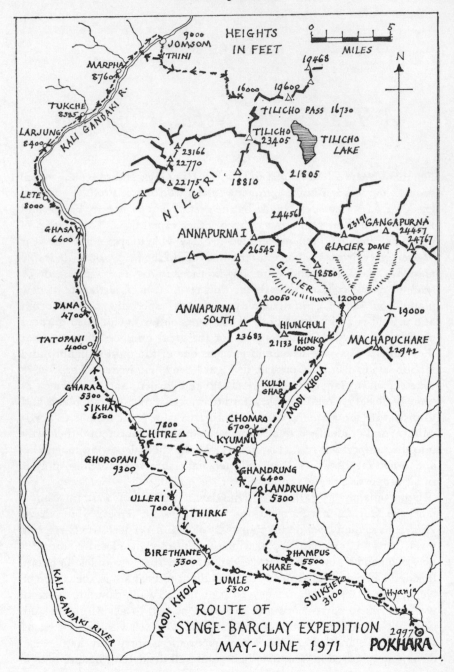

HEIGHTS
IN FEET

HEIGHTS
IN FEET

0 ․․․․ 5
MILES

N

JOMSOM 9000
THINI

MARPHA
8760

19468

19600
16000

TILICHO PASS 16730

TUKCHE
8585

TILICHO
23405

TILICHO
LAKE

LARJUNG
8400

23166
22770

21805

KALI GANDAKI R.

22175

18810

24456

23191 GANGAPURNA
24457

LETE
8000

N I L G I R I

ANNAPURNA I

26545

GLACIER DOME

24767

GHASA
6600

GLACIER

18580

12000

DANA
4700

ANNAPURNA
SOUTH

20050

19000

TATOPANI
4000

23683

HIUNCHULI
21133

HINKO
10000

MACHAPUCHARE
22942

KULDI
GHAR

MODI KHOLA

GHARA
5300

SIKHA
6500

CHITRE 7800

CHOMRO
6700

KYUMNU

GHOROPANI
9300

GHANDRUNG
6400

LANDRUNG
5300

ULLERI
7000

THIRKE

BIRETHANTE
3300

DHAMPUS
5500

KHARE

MODI KHOLA

LUMLE
5300

SUIKHET
3100

Hyanja

KALI GANDAKI RIVER

ROUTE OF
SYNGE-BARCLAY EXPEDITION
MAY-JUNE 1971

2997
POKHARA

permits, and send with you one or two experienced Sherpas who engage porters, arrange camps and look after food. Mountain Travel of Khatmandu looked after ours and did it very efficiently.

In 1971 I was able to make a trek together with Sir Colville Barclay, a plant-loving friend and a neighbour. We went up into two valleys that lead up to the great group of Annapurna and Dhaulagiri, two vast mountain masses which are among the highest in the Himalaya. We were able to leave Pokhara in central Nepal on May 21st and headed first for the Modi Khola valley which leads up to the Sanctuary of Annapurna, an area at the foot of the high snows and surrounded on three sides by mountains. On the east was Machapuchare, the Matterhorn of Nepal, only it is 10,000 feet higher. It was perhaps the most superb mountain that I have ever seen, a perfect pyramid as seen from Pokhara but a fish-tail at the summit as seen from the west side; its name means the Fish Tail. Very early in the morning, from Pokhara we got our first view of the mountains although they were still about fifty miles distant. They looked ethereal and remote, floating seemingly above a sea of cloud which later rose over them – It was in the early morning and again in the evening that we got our best views, for the monsoon was coming – even from this distance they seemed so much larger than any alpine peaks. Our first camp was at Hyanja, a village where the Tibetan refugees have a large settlement. They help to support themselves by knitting sweaters of yak wool and making rather heavy jewelry with lesser precious stones. They were a cheerful people and the children liked to come and see what we were doing: they were more round-faced and Mongolian than the Nepalese but the men were mostly taller and broader. They managed most of the horse trails, taking rice up to the higher villages and bringing down salt from Tibet.

The next three or four days were among our loveliest. There was an adequate path, first through the now largely dry rice fields scattered with little blue annual gentians, then up and over ridge after ridge, first through open scrub country then through the forest. In the mornings, from clearings or turns in the paths, with the sun shining it was fine enough to give us magnificent views of Annapurna. We quickly got into our usual routine, up a bit before six when Pasang, our chief Sherpa, called us, bringing for each a small basin of hot water and later an old teapot of coffee and some biscuits. The camp was packed up with great expertise by Pasang and the porters and we usually managed to start walking about half past six. We would stop for a combined breakfast and lunch about half past ten for an hour and a half, by a stream or a spring if we could find it, and would have a good meal, sometimes starting with porridge,

153

cooked by Tensing, our other Sherpa, while the porters cooked for themselves a good hot meal and had a long rest. Then we trekked on till about
three in the afternoon, when we made camp, had a short rest and some
tea and settled down to preparing our specimens and writing field notes.
About seven came dinner and soon after that we went to bed with candles
for a short read, but quickly blew them out. We were able to keep to this
routine, wet or fine, throughout the trek and found it a very good day,
quite long enough but we felt no great need to hurry and stopped for
plants or photographs or rests at quite frequent intervals. I hate being
hurried in the mountains and this was a holiday and we wanted to savour
the mountains and their wonderful scenery to the full.

In the lower zones there were great figs with aerial roots, one looking
rather like a poplar with long drip tips to the leaves. Always the same pair
were planted together symbolizing the male and the female, usually close
to a few huts or a village. Often the branches were full of orchids, *Dendrobiums* with pink flowers or *Coelogynes* with white and yellow ones. At about
5,000 feet we came into the zone of the giant *Schimas*, trees closely related
to the camellias with thick leathery leaves, red when young, and quite
large cup-shaped creamy-white flowers, each with a central boss of yellow
stamens. Some relations of these from farther east grow in Cornish
gardens, but I had seen none before so large as these handsome fifty-foot
trees. They are *Schima wallichii* and I do not know of any in English
gardens but they might prove too tender. Another very fine plant of this
area, also with magnificent scarlet foliage, was *Lyonia ovalifolia* which had
the appearance of a *Pieris*, in which genus I placed it at first. The flowers
were a good white, like larger lily of the valley bells and hung in large
clusters. Growing there on the forest edge it seemed as fine as any *Pieris* I
had seen. I only hope that some of our seedlings may prove hardy enough
for outdoor cultivation in this country but plants from this altitude are
expected to be on the borderline of tenderness.

It was a dream world as one trekked through the luxuriant forests that
cover the approaches to the big mountains. The variety of greens alone
made innumerable pictures, while many of the young leaves were red or
crimson with long pointed drip tips since the atmosphere is very humid.
The aroids between about 5,000 feet and 10,000 feet were particularly
exciting and grew abundantly in the rich humus and mosses of the forest
floor: the *Arisaemas* looking like cobras or like weird arums. They were
two or three feet high and had striped flower spathes and elongated central
and club-like spadices. The commonest was *Arisaema erubescens* like an
enlarged Jack in the Pulpit with a magnificent maroon and green striped
spathe, curling towards the tip. All green was *A. tortuosum* with a spadix

much longer than the spathe, and the poise of a large green bird just about to take off. But the largest of all was *A. griffithii* with a vast maroon spathe marked with light green veining, eight inches or more across and hanging down right over the spadix, an exotic almost a gross plant but exciting to find. The arisaemas are separated easily from the arums by the divided instead of entire leaves; and from *Dracunculus*, the great black dragon arum of the Mediterranean, by having several ovules in the ovary, a character which will not, however, help the amateur much if he doesn't want to cut up such a monster. But the most amusing of all was a little bright yellow aroid only about six inches tall which grew among the moss either as an epiphyte or on the ground. I had never seen one like it and was quite defeated for a name when we first saw it. However, our hopes of a new species were dashed later at Kew where it was identified as *Remusatia hookeriana*, a plant collected in Sikkim by Sir Joseph Hooker.

Sometimes one came to an opening and would look up to see the peak of Machapuchare or the great ridge of Annapurna floating sugar-white above the cloud and so far, apparently, above us that they too appeared unreal. Machapuchare was certainly one of the most beautiful mountains that I have ever seen with its symmetrical cone of snow, rock and ice. We wandered under great rhododendron trees, some fifty feet high, some with several trunks. These were the deep red *Rhododendron arboreum*. Most of them had finished flowering although an occasional truss remained. The trees were magnificent though in their young foliage, with the burnished silver under surface to the young leaves which stood up vertically at this time, so that they gleamed in the sunlight. Higher up the pink and white flowered forms grew, also *R. barbatum* and two dwarfer species, but central and western Nepal do not have the wealth and variety of rhododendron species that one finds farther east.

In clearings *Rosa moschata napaulensis* rambled luxuriantly. Often it was ten feet high and more across, a very decorative plant but excessively prickly. The flowers were in large clusters, strongly scented and creamy-white. It should be grown much more in English gardens and is available in some English catalogues, often listed under *R. brunonii* under which name it used to be grown. In England it will flower in mid-June and scramble up into an old tree. The best and hardiest form is known as 'La Mortola' and probably the clusters of flowers are slightly larger than in the form we saw. I ordered one for the garden on my return and have planted it near the foot of a dead willow and look forward to constant reminders of our mountain trek, when it covers the old tree. It should not take many years since after only half a year it is already up to five feet.

Another familiar climber was the white *Clematis montana* which we found growing in one place only and covering its host bushes. It was only a medium form, probably not so large in flower as some in cultivation, but still it was exciting to see it in the wild. Near it was a group of the rare *Rhododendron cowanianum* with bright purple saucer-shaped flowers, and we were able to find a little seed. It was a rather straggly bush and the flowers were somewhat small in comparison with those of many of the giants of the genus, but nevertheless I was glad to find it. In flower it was somewhat similar to the more common and more dwarf *R. lepidotum* but it had a better colour and flower than most forms and grew bigger. In this area between 8,000 and about 11,000 feet, we also saw our only spike in flower of the majestic giant lily now known as *Cardiocrinum giganteum*, although there were plenty of young plants with their big basal heart-shaped leaves and a few young spikes in bud. Probably its main flowering is during the monsoon period, just a little later. What a superb plant it is, a stem six or more feet tall with pure white long narrow trumpet-shaped flowers hanging down all round and each eight inches or more long. I know that we both had seen finer spikes of it flowering each year both in the woodland garden at Wisley and in the Savill Garden at Windsor Great Park, but this did not interfere with our excitement at seeing it towering out of a group of bushes in a little side valley with the great mountains above – its own habitat. May it long be preserved there! The only true lily which we saw was *L. nepalense* growing in pockets between damp mossy rocks or even in the moss and humus covering the rocks. We saw this in several places, but unfortunately the plants were all small and only a very few showed any signs of bearing just a single flower. But obviously its flowering season was appreciably later, probably about early July during the monsoon. It would then be a very dramatic plant, with large nodding flowers of deep jade green opening almost flat and splashed with deep crimson around the base. Unfortunately it is rarely seen thus in England and is probably a bit tender for gardens outside the milder areas, and obviously it requires plenty of moisture when in growth.

The forest floor was rich in plants as well as ferns and mosses of many kinds. There were dark brown and yellow *Calanthe* terrestrial orchids, probably *C. tricarinata* as well as other species; the spikes were a foot and a half tall with quite large flowers, the lip being deep brown and the petals and sepals a bright yellow, while the leaves were wide and pleated like a fan. Another species had chocolate and crimson flowers. We were able to bring back a few pseudobulbs as it was plentiful.

Another unusual orchid to find here was *Spiranthes australis*, a little green-white twisted spike only a few inches high and very close in general appearance to our wild Ladies Tresses, *Spiranthes spiralis*, which grows in some of the meadows in Sussex around our home and flowers in the late summer. This, however, grew only in more open spaces in short grass near the streams. The large Himalayan Herb Paris, *Paris polyphylla*, was a more occasional plant. This is larger than our native and the alpine species, and the petals are replaced by long narrow yellowish filiform streamers radiating from the centre of the flower. There were also six leaves in place of the four of our native species and it was a more decorative plant. There were several smaller Solomon's Seals with smaller hanging bells than in our common garden plant. These were *Polygonatum cirrhifolium* with little white bells and curling tendrils at the ends of the leaves, and *P. punctatum* with white lilac-spotted bells, and the more lilac-mauve *P. hookeri* which has only a single flower hanging below each leaf. Another plant close to this was *Ophiopogon intermedius* which looked like slightly larger lilies of the valley in flower and we found both white and pale mauve forms. It is easily distinguished though from the lily of the valley by its grass-like leaves. It was very pretty and should make an interesting plant for shadier and moister places in the rock garden. In distribution it stretches right along the mountains to China. Little violets, mostly *V. serpens*, also grew among the mosses.

Among the shrubs were honeysuckle such as *Lonicera myrtillus* with greenish-white flowers, a rather dull *Viburnum V. erubescens*, and higher up the more decorative *V. grandiflorum* with deep pink flowers although it was not so good a form as I had seen in cultivation in England. In gulleys up to about 8,000 feet there was the large *Mahonia nepaulensis*, with large leathery prickly leaves, larger than I had seen in English gardens but looking much more ragged and dishevelled. It had long finished flowering. On the lower slopes was a pretty pinkish-magenta flowered shrub with velvet pleated leaves. The flowers were saucer-shaped with a pronged trident of stigma and stamens and it obviously belonged to the *Melastomaceae* and was later identified as *Osbeckia nepalensis*. It grew below 6,000 feet and so is likely to be tender in English gardens. It is not far removed from a plant of the same family from Brazil, *Tibouchina semi-decandra*, which is commonly grown in cool greenhouses but the flowers of the *Osbeckia* cannot rival the glowing royal purple of the *Tibouchina*, a plant almost unrivalled for its colour. A little higher there were fine evergreen oaks such as *Q. lamellosa*, handsome large-leaved elders such as *A. nepalensis* and shrubs such as *Symplocos crataegoides* with conspicuous shining foliage, *Daphniphyllum himalayanum* with bright yellowish-green leaves and a *Cotoneaster* sp. close to our well known *C. microphyllus*. We saw only one specimen of *Magnolia campbellii*, and that on the divide between the Modi Khola and the Kali Gandaki valleys. The flowers were long over but a villager assured us that they had been white. The foliage was unmistakable. Probably it is commoner as one goes eastwards in the country towards Sikkim.

Primulas began in the forest area and on the banks by villages and clearings but below 9,000 feet they had mostly finished flowering. There was one in shady damp places which appeared to be of the Petiolarid section and I was sorry not to have seen it in flower, another on damp rocks seemed to belong to the Farinosae section and had a few pale pink flowers with a yellow eye and a small whitish ring round it. Around 9,500 and 10,000 feet, on damp mossy rocks where the forest was getting thinner, the primulas got more exciting. On one rock we found a small colony of *P. sharmae* a rare species probably not at present in cultivation. It was a very lovely flower, very dwarf, only three to four inches tall, bright violet-mauve with a pale yellow eye. The flowers were quite large, each being over an inch across, and in small clusters not unlike those of *P. tayloriana* from my recollection of the earlier flowering of this rare species from Tibet. Both belong to the Farinosae section. On another rock we found *Primula reidii*, pale blue-mauve and white forms growing together on the same rock. They probably both belong to the variety *williamsii* of which a white form has been described. This is one of the loveliest of primulas of the Soldanelloid section and the very graceful hanging bells do bear some resemblance to a *Soldanella*. In Nepal the spikes were about six inches tall with several flowers, some ivory-white the others pale blue-mauve, with rather open bells. The leaves are hairy and crinkled. The type of this was collected near the Kali Gandaki valley, the second one in which we trekked. Another beauty was *P. involucrata* but we saw that in better form and more plentifully later, westwards on our trek towards the Tilicho Lake and so we will discuss it under that area.

The track was mostly good, and above about 7,000 feet, where Chomro was the last village, we met hardly anyone except for one Nepalese party of hunters climbing to look for the magnificent pheasants of the area. We heard these calling sometimes but unfortunately never saw one flying, only a few dead birds hung in villager's houses. Their plumage is most magnificent and we feared that they were becoming much depleted through shooting. After one rather fearsome crossing of the raging Modi Khola, quite low between the villages of Landrung and Ghandrung, we kept to the west bank of the river. The bridge here consisted only of a single smoothed and horribly slippery mossy tree trunk well above the torrent below. The river was quite high since we had had rain most of the evening and night before. Our porters took it most nonchalantly and the Sherpas even were able to offer us a comforting hand. Everlastingly we seemed to be trekking up or down across the small ridges and valleys and only occasionally could one find a flat enough spot to pitch a tent or get any wide view.

At both Ghandrung and Chomro we stayed on the verandahs of houses in the village, and our Sherpas quickly rigged up an old tarpaulin to protect us a little from the eyes of the villagers. The children were delightful and some very good looking, but naturally they were curious as to what we might be doing and few understood why we should wish to put plants pressed flat between sheets of paper. The villagers have recently learnt the art of splitting some of the slaty stone which abound, and had made very fine large mottled slates for their roofs and used some for paving in the villages. Also on approaching the villages on either side, instead of the usual slippery slope with occasional logs as crosspieces one ascended by a really fine stone staircase, sometimes five hundred or more steps for half a mile or so. It was arduous and tiring but a great achievement in path making, the slabs often being three feet or more square. The houses look fine with these stone roofs but are still built without any chimneys so that they quickly fill up with smoke. They are also without any form of sanitation for the most part. However, there is plenty of hillside and most of the older people at any rate are discreet in their operations. All the people we encountered were friendly and hospitable if we asked to use the verandah of their houses for the night, but such matters were arranged through our Sherpas. A few of the villagers spoke some English but otherwise we had to communicate through our Sherpas. The Nepalese are a fine looking race, mostly rather small with delicate features and bones. They are not nearly so Mongolian as are the Tibetans. From an early age the children carry small loads on their backs, since this is their only form of transport. Often one will come on a little girl of six or seven carrying on her back a younger child tied round with an old rug.

As we got higher up the valley the vegetation changed, although gradually rather than suddenly as I had seen on other equatorial mountains. At about 9,000 feet one got out of the zone of the larger rhododendrons and deciduous trees and there were more bamboos until, above about 11,000 feet, there was a dwarf scrub with patches of meadow and long grassland. There were strangely few conifers on this western side of the valley although across the river we saw some giant silver firs and a few deodar cedars. The scenery was absolutely superb with the views of the mountains opening up all around. By one little side stream we saw our first *Meconopsis* rosettes at about 11,000 feet, lovely rusty furry leaves, but unfortunately the flowers were not yet out. The porters liked to strip down the spikes of buds and eat the centres like shoots of asparagus, regarding this as a delicacy. If many parties came up this would undoubtedly diminish the population of menconopsis, but there was an adequate number at present. On opening a fairly advanced bud the rolled petals showed that the flowers would open yellow and these were presumably *M. paniculata*. They were plants of the open areas rather than of the forest. There were also large clumps of an unusually fine spurge with large ochreous yellow flower heads and bracts. This was *Euphorbia wallichiana* and should make a very fine garden species. Only once have I seen it so used and that was in the Savill Garden of Windsor Great Park, and I much admired it then, but it was even finer in the wild up to three or four feet tall and often in clumps as much across. With it grew masses of the dwarfer *Anemone obtusiloba*, and both dark blue – and white-flowered forms grew intermingled with blooms each an inch or so across on one-foot stems like large blue or white buttercups.

The dwarf rhododendron of this area was *R. anthopogon*, a neat growing species about two foot tall, with small aromatic leaves which gave out their scent as one walked through them, and light trusses of rather small cream-coloured flowers. This is a very variable plant and in the Tilicho area we saw some with pink-flushed flowers. Among bushes and in open grassland we found the beautiful *Fritillaria cirrhosa*, a plant that I had particularly hoped to see. The bells are quite large, and chequered with deep crimson and yellowish-green on a whitish base while along the centre of the outside of each pointed petal is a green central stripe. Above each flower are three leaflets with curling tendrils at the tips. The bulbs are very small, only a little larger than a pea. In cultivation I expect that they will require different treatment from the Mediterranean and Middle East species, since after flowering they continue to have a very damp period during the monsoon and their resting period will be our winter when they are snow covered. In the Tilicho area we again found this fritillary but here the basic colour of the flower was greener and these were more purplish-crimson, particularly along the edges of the petals. Here it only seemed to grow in the protection of bushes where it could not be grazed. In both the Modi Khola and the Tilicho areas we were ahead of the flocks of goats and water buffalo which later are driven up for grazing, and this was a very great advantage.

Above 11,000 feet and up to the remaining patches of snow the chief primula was the well known *P. denticulata* with large globular heads of mauve or purple flowers, but we also saw some unusually blue ones. I was only sorry that our photographs of these did not succeed in showing their true colour, but they were certainly much more blue and so more attractive than any I had seen previously. They were abundant. Another primula which excited us very much was *P. strumosa*, a Petiolarid primula growing

PLANTS OF THE PACIFIC STATES

(*Above*) *Yucca whipplei* which has rapier-sharp leaves, in S. California.

(*Below*) *Phlox diffusa* by Crater Lake in Oregon. The flowers are mauvish blue.

THE NEPAL HIMALAYA
(*Left*) Looking across to Dhaulagiri with the village of Thini in the Kali Gandaki valley. (photo: Sir Colville Barclay).

(*Below*) The high ridge of Annapurna from the Modi Khola valley.

SOME FLOWERS OF NEPAL

(*Above left*) *Lyonia ovalifolia* in the Modi Khola valley. The flowers were like those of a very fine Pieris.

(*Above right*) *Arisaema griffithii*, a giant aroid in the Modi Khola valley.

(*Below*) When we reached camp we settled down to our herbarium work. On the author's knee is a lovely Dendrobium orchid, left Sir Colville Barclay.

SOME MORE FLOWERS
OF NEPAL

(*Top*) *Pleione humilis* growing
in the moss on tree trunks.
(photo: Sir Colville
Barclay).

(*Centre*) *Saussurea* species at
16,000 feet. on the Tilicho
Pass behind Annapurna.
This is probably *S. sacra*
which also grows in Kashmir
where it is known as Yogi's
King-Plant.

(*Bottom*) *Stellera
chamaejasme*, a Daphne like
plant and one of our most
choice finds on the Tilicho
Pass.

in moist places where the snow had only recently melted and bearing large heads of bright yellow flowers. It was a plant of great distinction and I am sorry that it is not in cultivation generally for it has been recorded from many parts of the Himalayas. The stems and petioles were covered with a white farina which offset the clear yellow of the flowers. Some of the flower heads were several inches across with a large number of flowers.

Unfortunately when we got to the edge of the Sanctuary area a storm came up and obscured the views and our Sherpa thought it best to go back a bit to a more sheltered area where a previous climbing expedition had made a base camp. It was a delightful spot and it was subsequently a great regret to us that we did not have sufficient days available to return the next day and camp there for two or three nights, when we could have explored much more thoroughly the flowers of the area. So reluctantly we turned back down the valley and below Chomro we crossed over the Deorali pass, an area of about 10,000 feet, westwards into the Kali Gandaki valley to a delightful village called Chitre. Our path lay through very damp moss forest and it rained nearly all the way; in places it was a ghostly forest with dark conifers and vast rhododendrons and everywhere streamers of damp moss. There were few flowers. The undergrowth was partly formed of *Daphne bholua* which at this season had black berries. It is inconspicuous in flower but is the shrub from the bark of which a local paper is made.

The valley of the Kali Gandaki is one of the old trade routes from Tibet, and so the track is better and wider in order that horse parties can negotiate it, taking up rice or grain and bringing back salt. These are now managed mainly by the Tibetan refugees and frequently we stood aside at an awkward corner to allow the tinkling horse or mule parties to pass, coloured cockaded plumes waving in the breeze from their heads. The Kali Gandaki was a very different kind of valley, deeply eroded, much drier and barer than the Modi Khola: so we were able to see two quite different forms of vegetation. The reason for the difference was that much of it lay north instead of south of the main Annapurna range so that the rain of the monsoon was partly expended against the high ranges before reaching the valley. In its lower reaches it ran between the two giant massives Annapurna and Dhaulagiri, each with many peaks over 25,000 feet. As it was a trade route the villages were slightly larger with paved streets through the centre and some excellent wooden carving on the houses. The Nepalese have obviously skilled craftsmen in this art. There were also more signs of Buddhist interest in rows of prayer wheels to be turned as one entered the villages and little temples and arches with decorative painted ceilings across the paths. To cross the river, which was quite wide, we had to descend down to below 4,000 feet at a place called Tatopani named after the hot springs there which bubbled up beside the river. However, there was a fine suspension bridge about fifty yards long and with a good hand rail on each side, the kind that one could cross without any qualms. The vegetation was much more tropical with bananas and ginger-like plants although mostly secondary scrub and the larger trees had been cut for timber and for wood. Strangely there were large *Crinums* with white spidery flowers growing outside the villages. It was sultry, hot and steamy, and we looked forward to getting to the higher levels again. At Tatopani we left some of the bags of plants collected in the Modi Khola in the shade of one of the headmen's houses. They were in plastic bags pierced with large air holes to prevent the plants becoming too damp, but this was not enough and we lost some which rotted off, a great disappointment. On another trip I think it would be better to plunge the plants in the earth and protect them with baskets or brushwood from damage by children, goats, water buffalo and other perils. Also they would have been better left at one of the higher villages where it was not so hot. Our Sherpa advised strongly against the risks of leaving them outside but I think nevertheless we should have tried it and taken the risk. We lost in this way our plants of several primulas which we had particularly wanted to keep.

From Tatopani up to Jomson, the nearest village to Tibet to which our trekking permit would allow us to go, was three to four days. The flora was very different to the other valley as we got up. There were more conifers, such as *Pinus wallichiana*, a lovely tree with long needles, and higher up some grey junipers. Planting around the villages was more luxuriant with small fields of sweet corn and potatoes and barley or millet. There were also large bamboos some thirty or more feet tall. In the pine woods were lovely clumps of the terrestrial *Calanthe* orchid, finer than we had seen in the previous valley. The *Rosa moschata* was here replaced by *Rosa sericea*, a very spiny shrub with clear pale yellow or creamy flowers. Much of the vegetation was xerophytic and adapted to the much drier conditions. The river bed widened out into a broad tableland of damp scree nearly a half mile across in which only a few plants grew, while the main river and numerous side streams rushed, grey with silt, in deep channels. This scree was fairly firm and formed the main route. The cavalcade of horses crossing the desolate expanse and the grey eroded steep hills around looked very like pictures I had seen of Tibet.

There were, however, a few interesting plants. Near here we saw our only plant of *Cornus capitata*, a small tree about twenty foot tall. It was certainly the finest species of *Cornus* that I had seen. The bracts surrounding the flowers were like petals and a cool lemon-yellow in colour so that the heads, four or more inches across, looked superb when outlined against the blue sky and snowy mountains, for we were lucky in a fine spell when we saw it. This is a plant well known in gardens in the really milder areas of Cornwall and the west but is unfortunately tender in most other areas. With it were shrubs like deutzias, indigoferas, philadelphus, jasmine and cotoneaster, genera all well known in English gardens, but we saw nothing to surpass those we already have. We also saw a few trees of the horse chestnut *Aesculus indica* in full flower, and the surprising contrast with the surrounding bareness made one realize what a fine flowering tree this is. In the undergrowth there were little yellow violas, *V. biflora* the perky little flower one sees in so many countries of Europe. Prickly shrubs became more prominent as one climbed higher into the drier zones. The finest at about 9,000 feet was *Sophora moorcroftiana* var. *napaulensis*. The foliage was silvery-grey and the flowers pale mauve and the whole effect was most attractive. This is a plant that would be well worth trying in English gardens in spite of its spines, but unfortunately there was no seed. Equally spinous were the *Caraganas*, relatives of the European gorses but some of these here had large pea-like flowers, creamy-yellow, heavily flushed with pink. There were even dwarf and spiny small shrubby honeysuckles with white flowers.

The wind up the valley from the south was extreme in the higher regions and seemed to blow most of the day. At Jomson a barrier across the path and a police post barred the way further north and our trekking permits were held there until we returned. We were glad, however, to be allowed to go eastwards towards the Tilicho yak pastures and lake and to climb again into rather less infertile country – an area which we found rich and interesting botanically. On the way up to the Tilicho we had most dramatic views of the Dhaulagiri peaks, bright and snow-covered, arising out of the clouds of the dark valley below with the white flat roofs of the little village of Tinne in the foreground. The track was much narrower and we hardly saw anyone after we left Tinne but the plants were of particular interest. *Stellera chamaejasme* that elusive relation of the Daphnes, which ranked high in the list of plants we hoped to find, grew in dry open patches among the rocks, great clumps two or three feet across and about eighteen inches tall of glaucous foliage, each stem ending in a cluster of pink buds and white flowers, sweetly scented. Some of the leaves and stems had turned pink also. The root of *Stellera* is like a large forked carrot and only a young plant could be attempted for collection since the soil was rock hard and baked around it. It was exciting to see this in such masses, real aristocrats of the plant world. Unfortunately they rarely set any seed, nor are they easily propagated from cuttings, and so it had almost died out of cultivation until we were able to take home a few plants. However, I did get a surprise this year when I saw a very large old clump in full flower on the rock garden of the Royal Botanic Garden, Edinburgh, and it must have been there for many years.

Along this trail we had one of our most lovely camps in an area known delightfully as the Nam Phu Yak pasture, and here at about 13,000 feet it was greener again and there was still some moisture in a little stream bed where grew *Primula involucrata*, of the Farinosae section, a very beautiful plant and flowering magnificently. Forms with white or pale pink flowers intermingled. The flowers are poised delicately in small clusters on slender stems, each flower opening flat with a pale yellow centre and bilobed petals; they are about an inch or a little more across. It only seemed to grow in damp places and with it grew *Primula sikkimensis*, but most of these were still in bud.

Dwarf blue irises, *I. kamaonensis*, flowered all round in the short grass and grew in great abundance. The flowers were variable in colour from a light blue-mauve to a deep purplish-mauve, all heavily blotched with deeper colour and with a golden beard. It was rather like a dwarf tall bearded iris but only about six inches tall. We even found a few white forms with slaty-blue markings and I have seen no other record of these. One of the blue forms flowered in a pot for me this summer and I hope to try to establish some outside, where they should be quite hardy. It flowered so freely that there were great patches of colour on the ground. Among the iris were a few with much flatter flowers and these turned out to be *Iris goniocarpa*, a Tibetan species and probably a new record for Nepal.

Grazing gently on the tussocks of grass we saw our first yaks, like great shaggy mounds, even larger than Highland cattle though without their vast horns. With their grass they probably consumed vast quantities of *Primula denticulata* which was also abundant in this area.

Another plant that had an unrivalled grace was *Paraquilega microphylla*, a plant of north-facing shady camp rock cracks and only rarely seen. It forms tussocks from a matted rather prickly base, and has small alpine poppy-like flowers drooping gently and of the palest mauve set among very finely divided foliage like a bluish-grey maidenhair frond. We only found it at one place, another demonstration of how very local many of these Himalayan plants are. It has, however, been recorded fairly freely from Kashmir while there is another allied species in western China, *P. anemonoides*, which has slightly larger flowers.

Rhododendron anthopogon, although here with slightly pink-tinged flowers, and a *Caragana*, a very prickly gorse-like dwarf shrub with quite large creamy flowers, also tipped with pink and pea-like, formed the main scrub but there were also prickly dwarf honeysuckles with small white sweetly scented flowers and grey leaflets. There were also dwarf anemones, pink starlike forms probably of *A. narcissiflora*, although only a few inches tall, and the very lovely creamy-white rather cup-shaped species, almost sessile on the ground. It was probably *Anemone demissa* and it should certainly make a very good rock garden plant. In appearance it was not unlike a larger and much more dwarf *A. rupicola*, a species which has also been recorded from these regions although we did not see it. Another exciting plant was *Thermopsis barbata* about a foot and a half high, with silvery and furry much divided foliage and flowers so dark that they were almost black. They were pea-like, quite large and borne in clusters like those of a lupin. The dark maroon colour was relieved at the base of the standards with a brighter almost orange-reddish flush which lightened up the flowers. This is another plant that I would like to grow but unfortunately there was no seed.

As we got higher the trail disappeared and we made our way along the ridges up to just over 16,000 feet, which seemed high enough for two sixty-year-olds. It was much more barren here with rocks and screes and shale, and one felt on top of the world since it was a wide ridge, but at this height we walked more slowly and stopped for breath much more frequently. There were plants appropriate to the region: dwarf cushion-like androsaces, encrusted saxifrages both white and yellow, and sedums as well as edelweiss species, while the rocks were stained with great patches of orange and grey lichens. The saxifrages mostly had small encrusted rosettes and formed large tight cushions with whitish flowers on inch-long pale pink stems, while the sedums, probably *S. humile*, formed tussocks covered with bright rather acrid yellow flowers. There was also a little *Trollius*, probably *T. pumilus*, with dwarf yellow flowers. But by far the most exciting of our finds in this zone was a *Saussurea*, a member of the Compositae, looking like a small white globe since the whole plant body is enclosed in long shaggy cobweb-like silvery-white hairs, while the flowers are hidden deep inside the globe and were invisible, although somewhere in the rosette there would have been a small hole for a pollinating insect to enter. Dewdrops gleamed silver among the hairs. Unfortunately we do not yet have an identification of it, but it was different from the more commonly recorded *S. gossipiphora.* The little androsaces mostly had rounded rosettes, also covered in cobweb like the rosettes of the alpine *Sempervivum arachnoideum* but smaller, while the flowers were deep pink with a yellow eye in a little cluster on stems only an inch high, a fascinating little plant for the alpine house. It is probably *A. muscoidea* var. *longiscapa*.

When we returned from the higher levels where we had only taken Pasang, our head Sherpa, and two well clothed porters, Tensing our Mountain Travel cook was ready with a choice of tea, cocoa or coffee, all in enamel pots on an improvised tray, very welcome after a long day, but still a most exciting one. Our return downwards was naturally much quicker and easier than our climb upwards and we were able to combine two days' upward trek into one without becoming too tired. The monsoon clouds were gathering but sometimes we were able to enjoy most superb views of the peaks emerging from the billowing clouds, dark and stormy below and silvery white above, while sometimes the light would come in a beam through the clouds to brighten a small patch of snow or ice. As we got lower the rain increased until it was almost continuous. Nevertheless, there were compensations such as the masses of the orchid *Pleione humilis* around Goropani at about 9,000 feet, growing in the moss on the tree trunks and fallen trees. In one area they were very plentiful, flowering right up fifty feet of mossy trunk. The flowers are rather like those of the more commonly and probably more easily grown *P. formosana*, each two to three inches across with pale pinkish-purple side petals, and prominent deep yellow markings on the fringed lip, a very beautiful and fascinating flower. Other orchids were plentiful also in the lower areas: a pretty little Vanda with yellowish-green petals and a lip with deep crimson markings growing in the moss of a tree trunk, several *Dendrobiums* some with white flowers tipped with purple and yellow throats, *Coelogynes* of white, green and yellow and a single *Cymbidium* out of flower. Other epiphytes were a fine *Hoya* with pendulous branches and clusters of scented white waxy flowers with crimson centres, probably a form of *H. bella*. Also epiphytic was *Vaccinium retusum*, a pretty species with fresh pink and white flowers and wine-coloured buds. But the most tantalizing were some very large white flowers of a rhododendron which I found fallen beside the path. Most of these large-flowered rhododendrons are epiphytes, but a quarter of an hour's search upwards among the tree branches did not reveal the plant and unfortunately Pasang with my field glasses had gone ahead. The flowers seemed to me to be those of *Rhododendron nuttallii*, a species with the largest flowers of any northern Asiatic species, with corollas four inches or more long, yellow at the throat. I was particularly sorry not to have been able to find the plant, since as far as I know this species had not previously been recorded so far west in the Himalaya. We did not see *R. dalhousiae*, the species of this series which I had hoped to find in that area. Another exciting plant of the lower valley was a species of *Pandanus*, the Screw Pine with long palm-like foliage and large prickly cone-like fruits.

The rivers were now much more swollen and sometimes we had to wade across, while the paths along the ricefields had become thick wet mud into which one sank if one paused. The Nepalese were working hard in them planting out their rice seedlings in the rain, and the irrigation channels leading water into each patch were hardly needed. And so, very wet and muddy, we completed our trek along the road through Pokhara to the little thatched hotel by the airfield where we waited for our plane somewhat anxiously, since owing to the weather no plane had landed for several days and there was a long backlog of passengers waiting, including the members of an Austrian mountaineering expedition which had been climbing Dhaulagiri. However, luckily the weather got clearer and the plane came on the day for which we had a firm booking and we were able to get off to Khatmandu as planned, although leaving Pasang and Tensing behind to sort stores and come on by a later plane. It had been a wonderful and most rewarding trip.

Index

Prickly Pear, South American, 57
Primulas, 23, 95, 158
  P. allionii, 28–9, 31
  P. auricula, 30
  P. auriculata, 95, 99
  P. denticulata, 160, 164
  P. elatior, 43
  P. elatior var. *columnae*, 95
  P. farinosa, 18, 95, 99
  P. involucrata, 158, 164
  P. longifolia, 95
  P. marginata, 29
  P. marginata 'Linda Pope', 29
  P. minima, 30–1
  P. pedemontana, 23
  P. reidii, 158
    williamsii, 158
  P. sharmae, 158, Ill. between pp. 112–13
  P. sikkimensis, 164
  P. strumosa, 1960–1
  P. tayloriana, 158
  P. viscosa, 23
Pring, George, 150
Prunus prostrata, 46, 99
Psoraleas, 59
  P. bituminosa, 59
Pulsatilla, 16, 17–18, 32
  P. alpina subsp., 17–18, 32, 36
  P. alpina subsp., *alpina apiifolia*, 17, Ill. between pp. 112–13
  P. occidentalis, 143
  P. sulphurea, 17
  P. vernalis, 17–18, 143
Puschkinias, 96
  P. scilloides, 95
Puya, 128
Pyrenees, 15, 19, 24, 35–8, 91
Pyrola rotundifolia, 31–2

Quercus alnifolia, 80–1
  Q. castaneifolia, 99
  Q. coccifera, 54
  Q. lamellosa, 157

Rafflesia, 58, 115
Ramonda myconi, 37–8
  R. pyrenaica, 37
Ranunculus, 52, 67, 71–2
  R. acetosellifolius, 43
  R. amplexicaulis, 40
  R. asiaticus, 71–2, 75, 78, Ill. opp. p. 32
  R. cadmicus, 79
  R. glacialis, 19
  R. pyrenaeus, 20, 40
Remusatia hookeriana, 155
Rhodes, 51–3, 58, 67, 72, 75–8
Rhododendron, 31, 116, 120, 146, 159, 161
  R. albiflorum, 144

  R. anthopogon, 160, 164–5
  R. arboreum, 155
  R. barbatum, 155
  R. brookeanum, 120
  R. caucasicum, 92
  R. chamaecistus, 30
  R. cowanianum, 156, Ill. between pp. 112–13
  R. crassifolium, 120
  R. dalhousiae, 166
  R. durionifolium, 120
  R. ferrugineum, 26–7, 33
  R. hirsutum, 26–7, 33
  R. javanicum, 120
  R. lepidotum, 156
  R. luteum, 87
  R. maddenii, 149
  R. nuttallii, 166
  R. ponticum, 88
Rhodothamnus, 31, 33
Richards, Paul, 108 *ff*
Robinias, 59
Rosa brunonii, 155
  R. foetida, 41, 98
  R. 'La Mortola', 155
  R. moschata, 162
  R. moschata napaulensis, 155
  R. persica, 97
  R. sericea, 162

Sage, Jerusalem, 55, 81
Salvias, 102
  S. pratensis, 23
Sarawak, 106–24
Sardinia, 44–5
Satureia thymbra, 54
Saussurea, 165
  S. gossipiphora, 165
  S. sancta, Ill. opp. p. 161
Saxifrages, 16, 24–6, 165
  Porphyrion, 26
  S. aizodes, 26, 29
  S. aizoon, 25–6, 29
  S. lantoscana, 29
  S. lingulata, 26, 29
  S. longifolia, 26, 29, 38
  S. oppositifolia, 23, 26
  S. retusa, 26
Schimas, 154
  S. wallichii, 154
Schinus molle, 58
Scilla bifolia, 63
  S. italica, 40
  S. lilio-hyacinthus, 40
  S. sibrica var. *taurica*, 92
Scolymus hispanicus, 76
Scutellarias, 26
Sedum, 137, 165